SPACEMAN

An Astronaut's Unlikely Journey to
Unlock the Secrets of the Universe

Mike Massimino

CROWN
ARCHETYPE

Published in the United States by Crown Archetype, an imprint of the Crown Publishing Group, a division of Penguin Random House LLC, New York.
crownpublishing.com

Crown Archetype and colophon is a registered trademark of Penguin Random House LLC.

Photography credits appear on page 317.

Library of Congress Cataloging-in-Publication Data
Names: Massimino, Mike, 1962– author.
Title: Spaceman : an astronaut's unlikely journey to unlock the secrets of the universe / Mike Massimino.
Description: New York : Crown [2016]
Identifiers: LCCN 2016011667| ISBN 9781101903544 (hardcover) | ISBN 9781101903551 (ebook)
Subjects: LCSH: Massimino, Mike, 1962– | Astronauts—United States—Biography. | Hubble Space Telescope (Spacecraft)—Maintenance and repair—History. | Space flights—United States—History.
Classification: LCC TL789.85.M324 A3 2016 | DDC 629.450092 [B]—dc23
LC record available at https://lccn.loc.gov/2016011667

Hardcover ISBN 978-1-101-90354-4
Ebook ISBN 978-1-101-90355-1
International Edition ISBN 978-0-451-49727-7

Printed in the United States of America

Art for frontispiece and part openers is courtesy of Shutterstock/Grisha Bruev
Jacket design by Christopher Brand
Jacket photograph by Chad Griffith

10 9 8 7 6 5 4 3 2 1

First Edition

To Gabby and Daniel:

Thank you for showing me a love that I never knew was possible,

and for providing me with not only the inspiration

to follow my dreams but with the drive to set

an example for you to do the same.

CONTENTS

PROLOGUE

A Science Fiction Monster

On March 1, 2002, I left Earth for the first time. I got on board the space shuttle *Columbia* and I blasted 350 miles into orbit. It was a big day, a day I'd been dreaming about since I was seven years old, a day I'd been training for nonstop since NASA had accepted me into the astronaut program six years earlier. But even with all that waiting and planning, I still wasn't ready. Nothing you do on this planet can ever truly prepare you for what it means to leave it.

Our flight, STS-109, was a servicing mission for the Hubble Space Telescope. We were a crew of seven, five veterans and two rookies, me and my buddy Duane Carey, an Air Force guy. We called him Digger. Every astronaut gets an astronaut nickname. Because of my name and because I'm six feet three inches, everybody called me Mass.

Ours was going to be a night launch. At three in the morning, we walked out of crew quarters at Kennedy Space Center to where the astro van was waiting to take us out to the launchpad. This was only the second shuttle mission since the terrorist attacks of 9/11, and there were helicopters circling overhead and a team of SWAT guys standing guard with the biggest assault rifles I'd ever seen. Launches had always had tight security, but now it was even more so. Digger was standing right next to me. "Wow," he said, "look at the security. Maybe it's a 9/11 thing."

I said, "I don't know. I think they're here to make sure we actually get on."

I was starting to get nervous. What had I signed up for? I could swear that one of the SWAT guys was staring at me—not at potential terrorists, but right at me. It felt like his eyes were saying, *Don't even think about running for it, buddy. It's too late now. You volunteered for this. Now get on my bus.*

We got on and rode out to the launchpad, everything pitch-black all around us. The only light on the horizon was the shuttle itself, which got bigger and bigger as we approached, the orbiter and the two solid rocket boosters on each side of that massive rust-orange fuel tank, the whole thing lit up from below with floodlights.

The driver pulled up to the launchpad, let us out, then turned and high-tailed it out of the blast zone. The seven of us stood there, craning our necks, looking up at this gigantic spaceship towering seventeen stories high above the mobile launcher platform. I'd been out to the shuttle plenty of times for training, running drills. But the times I'd been near it, there was never any gas in the tank, the liquid oxygen and liquid hydrogen that make rocket fuel. They don't put it in until the night before, because once you add rocket fuel it turns into a bomb.

The shuttle was making these ungodly sounds. I could hear the fuel pumps working, steam hissing, metal groaning and twisting under the extreme cold of the fuel, which is hundreds of degrees below zero. Rocket fuel burns off at very low temperatures, sending huge billows of smoke pouring out. Standing there, looking up, I could feel the power of this thing. It looked like a beast waiting there for us.

The full realization of what we were about to do was starting to dawn on me. The veterans, the guys who'd flown before, they were in front of me, high-fiving each other, getting excited. I stared at them like *Are they insane? Don't they see we're about to strap ourselves to a bomb that's going to blow us hundreds of miles into the sky?*

I need to talk to Digger, I thought. *Digger's a rookie like me, but he flew F-16 fighter jets in the Gulf War. He's not afraid of anything. He'll make me feel better.* I turned to him, and he was staring up at this thing with his jaw hanging down, his eyes wide open. It was like he was in a trance. He looked the way I felt. I said, "Digger."

No response.

"Digger!"

No response.

"Digger!"

He shook himself out of it. Then he turned to me. He was white as a ghost.

People always ask me if I was ever scared going into space. At that moment, yes, I was scared. Up to that point I'd been too excited and too busy training to let myself get scared, but out there at the launchpad it hit me: Maybe this wasn't such a good idea. This was really dumb. Why did I do this? But at that point there was no turning back.

When you're getting ready to launch, you have this big rush of adrenaline, but at the same time the whole process is drawn out and tedious. From the bottom of the launch tower, you take an elevator up to the launch platform at ninety feet. You make one last pit stop at a bathroom up there—the Last Toilet on Earth, they call it—and then you wait. One at a time, the ground crew takes each astronaut across the orbiter access arm, the gangway between the tower and the shuttle itself. You can be out on the platform for a while, waiting for your turn. Finally they come and get you, taking you across the arm into a small white room where they help you put on your parachute harness. Then you wave good-bye to your family on the closed-circuit camera and go in through the shuttle hatch. You enter on the mid-deck, where the crew's living quarters are. Up a small ladder is the flight deck. Neither is very big; it's pretty cozy inside the shuttle. Four astronauts, including the pilot and commander, sit on the flight deck for launch. They get windows. The remaining three sit on the mid-deck.

Once you're inside, the ground crew straps you in. They help you affix your helmet to your orange launch and entry suit. You check your oxygen, check your gear. Then you lie there. If you're on the mid-deck like I was there aren't any windows, so there's nothing to look at but a wall of lockers. You're there for a few hours waiting for everything to check out. You chat with your crewmates and you wait. Maybe play a game of tic-tac-toe on your kneeboard. You're thinking that you're going to launch, but you can't be sure. NASA's Launch Control Center will cancel a flight right up to the last minute because of bad weather or anything questionable with the spaceship, so you never really know until liftoff. Once it's down to about an hour, you glance around at your buddies like, *Okay, looks like this might actually happen.* Then it gets down to thirty minutes. Then ten minutes. Then one minute. Then it gets serious.

With a few seconds left, the auxiliary power units start. The beast that terrified you out on the launchpad? Now that beast is waking up. At six seconds you feel the rumble of the main engines lighting. The whole stack lurches forward for a moment. Then at zero it tilts back upright again and that's when the solid rocket boosters light and that's when you *go.* There's no question that you're moving. It's not like *Oh, did we leave yet?* No. It's *bang!* and you're gone. You're going 100 miles an hour before you clear the tower. You accelerate from 0 to 17,500 miles an hour in eight and a half minutes.

It was unreal. I felt like some giant science fiction monster had reached down and grabbed me by the chest and was hurling me up and up and there was nothing I could do about it. Right after we launched, I realized that all the training we'd had on what to do if something went wrong during launch—how to bail out, how to operate the parachutes, how to make an emergency landing—I realized that all those years of training were completely pointless. It was just filler to make us feel okay about climbing into this thing. Because if it's going down, it's going *down*. It's either going to be a good day or it's going to be a bad day, and there is no in-between. There are emergency placards and safety signs all over the interior of the shuttle, telling you what to do and where to go. That stuff is there to give you something to read before you die.

After about a minute, once the initial shock passed, this feeling came over me. I had a sensation of leaving. Like, really leaving. Not just good-bye but *adios*. I'd been away from home before, on vacations and road trips, flying out to California, going camping in East Texas. But this time, my home, this safe haven I'd known my whole life, I was leaving it behind in a way that I never had before. That's what it felt like: truly leaving home for the first time.

It takes eight and a half minutes to make it into orbit. Eight and a half minutes is a long time to sit and wonder if today is going to be the day you get it. You can't say much because your mic is live and you don't want to get on the comm and say anything stupid that might distract people. It's not the time to try to be clever. You just keep lying there, looking at your buddies, listening to the deafening roar of the engines, feeling the shuttle shake and shudder as it fights to break out of the Earth's atmosphere. You get up to three g's for about two and a half minutes at the end and you feel like you weigh three times your body weight. It's like you have a pile of bricks on your chest. The whole thing can be summed up as controlled violence, the greatest display of power and speed ever created by humans.

As you're leaving the Earth's atmosphere, the bolts holding you to the fuel tank blow. You hear these two muffled explosions through the walls of the shuttle—*fump! fump!*—and then the fuel tank is gone and the engines cut and the whole thing is over as abruptly as it began. The roar stops, the shuddering stops, and it's dead quiet. All you hear are the cooling fans from some of the equipment gently whirring in the background. Everything around you is eerily, perfectly still.

You're in space.

Once the engines cut and you're in orbit, the shuttle's no longer accelerating. Your perception is that you've come to a complete stop. You're moving at 17,500 miles per hour, but your inner ear is telling your brain that you're perfectly still; your vestibular system works on gravity, and without any gravity signals coming in, the system thinks you're not moving. So you have this sensation like you're lurching forward but then you come to a stop when the engines cut. You feel like you're sitting straight up in a dining room chair, except that you're still strapped down flat on your back. It's completely disorienting.

The first thing I did was ask myself, *Am I still alive?* It took me a moment to answer. *Yes, I'm still alive.* We'd made it, safely. It took me a minute or two to get my bearings. Then, once I felt acclimated, it was time to go to work. I reached up and took my helmet off and—just like I'd watched Tom Hanks do in *Apollo 13*—I held it out and let it go and floated it in the air in front of me, weightless.

part

1

I Want to Grow Up
to Be Spider-Man

A PERFECT GOOD

Your first week as an astronaut is a lot like the first week at any other job. You go to meetings, fill out paperwork, find out what's covered by the new health plan. In my first week, my astronaut classmates and I got lucky. As we were starting out, an astronaut reunion was being held at the Lyndon B. Johnson Space Center. All the living legends from the Mercury and Gemini and Apollo programs would be there, including Neil Armstrong, the first man to walk on the moon. My hero. Everyone's hero.

Our training manager, Paige Maultsby, was the mother hen taking us new kids through the orientation, and she put in a request asking if Neil Armstrong would come and speak to our class. He said that he would, but that he would only speak to us, the new astronauts; he didn't want some big public event with tons of people showing up.

I'd seen Neil Armstrong once before. In 1989, during graduate school, I'd worked as an intern at the Marshall Space Flight Center in Huntsville, Alabama, and that summer there was a big twentieth-anniversary celebration of the moon landing. Armstrong and the other crew members, Buzz Aldrin and Michael Collins, were there. From the back of a room full of thousands of people, I saw him give a speech, but I didn't get to meet him or shake his hand. Now, seven years later, not only was I going to get to meet him, but I was meeting him *as* an astronaut. This was the coolest thing ever.

Except I wasn't an astronaut yet, technically. When you first join NASA, you're an astronaut candidate, an ASCAN. For Armstrong's talk, the ASCANs gathered in the astronaut conference room, Room 6600 in Building 4S. It's a very important room. Every NASA flight gets its own patch, commemorating the mission with the names of the astronauts who flew it. On the walls of that conference room there's a plaque of every patch of every mission going back to Alan Shepard's first Mercury flight in 1961. You can feel the history of the place when you walk in. The goal of every astronaut who comes through is to get your name on that wall before you leave. We clustered around the conference table like eager schoolkids, and Armstrong came in and spoke for a few minutes. He was older but not ancient, thinning hair, glasses, suit and tie. He seemed warm and approachable, but at the same time he was someone you'd only approach with the utmost respect. When he got up to speak he was very soft-spoken, almost shy.

He talked to us for about fifteen minutes, and the whole time he didn't say anything about walking on the moon, not one word about being an astronaut—nothing. Instead he talked about his days as a test pilot at Edwards Air Force Base in California, flying the X-15, the hypersonic rocket plane that set speed and altitude records in the 1960s by flying 50 miles above the Earth, the outer limit of the atmosphere, the edge of space. That was how Neil Armstrong thought of himself: as a pilot. Not as the first man to walk on the moon, but as a guy who loved to fly cool planes and was grateful for the opportunity to have done it.

By focusing on his test pilot days and not on the moon landing, I think he was trying to tell us that life is not about achieving one great thing, because once that thing is over, life keeps going. What motivates you then? The important thing is having a passion, something you love doing, and the greatest joy in the world is that you get to wake up every day and do it. For him it was flying. He said, "Yeah, I got to fly to the moon, but I also got to fly the X-15." Just the fact that he got to go out and fly those planes every day, that's what had made him the happiest.

After he spoke, he took a few questions and said he'd sign pictures for us. He stood at the front of the conference table, and we lined up to shake his hand and get his autograph. I was toward the back of the line, and I noticed as I got to the front that every single person, when they got to the end of the table, did the exact same thing: They told him where they were when they watched him walk on the moon. I was thirty-three when I became an astronaut, one of the youngest in the class, which meant everyone in line was old enough to remember the moon landing, and everybody had a story to tell him: "I was at my girlfriend's house." "I was in my parents' basement." "I was in the Catskills." On and on and on. Because everyone on Earth knows where Armstrong was on July 20, 1969, so why not tell him where you were that day? I realized that this had been the man's whole life for the past twenty-seven years. Every day. Every new face he meets, they tell him the exact same thing, and he listens politely and nods and smiles.

I decided I was going to do something different. When I got to the front and my turn came, instead of telling him my moon landing story, I shook his hand and I said, "So is this what happens every time you meet people? They tell you where they were when you walked on the moon?"

"Yeah."

"You get it a lot?"

"Yeah, all the time."

"Does it ever bother you?"

He shrugged. "Nah, it's okay."

I never told Neil Armstrong where I was when he walked on the moon; I didn't want to, even if he said it was okay. But I remember where I was exactly, because it was the moment that changed my life. I was six years old, about to turn seven, sitting around the black-and-white TV in our living room with my parents and my sister, Franny, who was thirteen. She was wrapped up in her pink robe and I was in these baseball pajamas with pinstripes, worn and faded hand-me-downs I'd gotten from my brother. My mother's parents lived upstairs, and they came down to watch the moon landing with us.

I was glued to the television. Watching Neil Armstrong take those first lunar steps completely blew my mind. But seeing it on TV almost made it seem normal, like it could have been any old TV show. Going outside afterward made me think about how incredible it was. I remember standing in my front yard and staring up at the moon for the longest time, thinking, *Wow, there are people up there, walking around.* To a six-year-old kid in the suburbs on Long Island, it was the most awe-inspiring thing in the world. Something about it grabbed me down deep in my soul.

Going to the moon was a perfect moment, for me and for the

entire country. Life doesn't give you many of those. Everybody loved the Apollo astronauts—loved them. My father, my mother, my sister, my friends, my teachers. Nobody, no public figure, gets that kind of absolute, universal admiration. Especially back then. It was the end of the 1960s and everything was going crazy. You had people getting shot. Martin Luther King Jr. and Bobby Kennedy had been assassinated. Vietnam was tearing the country apart. Riots were breaking out every summer. And in the middle of all that, on one night the whole world stopped and watched and shared this one thing, this perfect good.

Even at that age I remember thinking, *This is the most important thing that's happening right now—and not just now, ever. This is going to mark our time on the planet: the fact that we were the first people to leave it.* Neil Armstrong and Buzz Aldrin and Michael Collins, they were space explorers. People were going to read about them five hundred years from now the same way we read about Christopher Columbus today. Those men became my heroes. They were the epitome of cool.

I turned seven in 1969, and there's something about that age that makes it a formative year in one's life. Two things happened for me that year: *Apollo 11* landed on the moon and—even more improbable than that—the Mets won the 1969 World Series. Space and Major League Baseball became my two greatest passions. The Mets' ace pitcher, Tom Seaver, was right behind my father and the *Apollo 11* astronauts on my list of childhood heroes. But on that night of the moon landing, the World Series was still months away. On that night I said to myself, *Nothing else matters. This is it. This is who I want to be.* Being an astronaut wasn't just the coolest thing ever, it was the most important thing you could choose to do with your life.

From that moment on, I became obsessed with space in the way that only a young boy can become obsessed. It was all I talked about. At my school's summer recreation program, we had a space parade in honor of the moon landing. The kids were dressed up in space-related costumes. I wanted to go as an astronaut. My mother was a seamstress. She took a gray elephant costume she'd made for me when I was in the first-grade play, cut off the tail, and added some of my dad's Army medals and an American flag on the left arm. We traded in the cardboard elephant ears for a plastic Steve Canyon jet helmet, added safety goggles, and we had my astronaut costume.

My brother, Joe, was working in downtown Manhattan that summer, and one day on his break he went to FAO Schwarz and got me this Astronaut Snoopy toy. It was about eight inches high, decked out in a full Apollo space suit: helmet, life support system, moon boots, the whole thing. I still remember watching Joe walking home from the bus stop with this Snoopy box in his hands. I opened it up right there in the driveway. That whole summer, all I did was play spaceman in the backyard, running around in my costume my mom had made with my Astronaut Snoopy. I played with that little guy until his enamel was cracked and worn and one of his legs broke off. (I still have him, only now he's been to space for real.)

I was obsessed with learning more about astronauts. The public library was right around the corner on Lincoln Road, and I'd go over in the afternoons and read anything I could find about the space program. They didn't have much, but whatever they had I probably checked out and read four or five times. They had a book about the original Mercury Seven astronauts called *We Seven*, and another about Gus Grissom, who'd died in the fire that killed the crew of *Apollo 1* on the launchpad at the Kennedy Space Center. I read *Time* magazine, *Life* magazine, whatever they had, whatever I could get my hands on.

That fall I started second grade, and all I talked about at school was space. I'd become this total space expert. My best friend back then—and still to this day—was Mike Quarequio, whom we called Q. He remembers me showing up for second grade and walking into class talking about spacewalking suits, the cooling systems they used, how the life support worked. I was known as "the boy in class who knows the most about space." I knew who the astronauts were, which kinds of rockets were used on which flights. I knew everything about space that a seven-year-old kid on Long Island could possibly know.

Even though I was obsessed with space, I never got into Flash Gordon or Buck Rogers or any of that. Space colonies and multiple dimensions and flying around with rocket packs—it was too far-fetched. What I loved was the science fiction of Jules Verne novels, like *Journey to the Center of the Earth*, *20,000 Leagues Under the Sea*, and *From the Earth to the Moon*. The thing about Jules Verne's stories was that he made them feel real. It was science fiction, but you felt like it was plausible, like it was set in the real world. In *Journey to the Center of the Earth*, they're digging their way down with pickaxes and shovels. In *From the Earth to the Moon*, a lot of what Verne predicted about space travel was accurate, from the type of metal they used to build the spaceship to the way they launched with the rotation of the planet to gain extra speed. And he was imagining all of it back in 1865!

I wasn't interested in the fantasy of space travel. I was interested in the reality of space travel. I was interested in how people got to space here and now, and at that point the only way to get to space was to join NASA, get an American flag on your left shoulder, and strap yourself into a Saturn V rocket. I only had one problem: Where I came from, kids didn't grow up to be astronauts.

A lot of people, when they meet me, can't believe I've been to space. They say I look like a guy who'd be working at a deli in Brooklyn, handing out cold cuts. My grandparents were Italian immigrants. My father's father, Joseph Massimino, was from Linguaglossa, near Mount Etna in Sicily, and he came over in 1902 to New York City and ended up buying a farm upstate in a town called Warwick, which is where my father, Mario Massimino, grew up. When my dad left the farm he moved back to the city, to the Bronx, where he met my mom, Vincenza Gianferrara. Her family was from Palermo, also in Sicily, and they lived in Carroll Gardens, an Italian neighborhood in Brooklyn. She and my dad got married in 1951. He was twenty-eight, she was twenty-five, which was pretty late in those days.

Although my dad never went to college, while he was working he started taking fire safety courses at NYU and eventually became an inspector for the New York City Fire Department. His job was fire prevention. He would go into apartment buildings and businesses and make sure they had the right number of extinguishers and sprinkler systems and safety exits. He was a smart guy who did a good job and kept moving up to eventually become the chief of fire prevention for the entire New York City Fire Department. My mom stayed home with us kids, for which she deserves a medal.

My parents lived in the Bronx, which is where my older brother and sister were born, but soon after that my parents decided to leave the city. They bought a house at 32 Commonwealth Street in

Franklin Square, Long Island, which is where I came along. I was born on August 19, 1962. My brother was ten years older than me and three years older than our sister. I was the mistake—or, as my mother would say more lovingly, "the surprise." She used to tell me that I must have come for a reason because she thought she was done having kids after my brother and sister.

Franklin Square is right outside Queens on Hempstead Turnpike. When I was growing up, the neighborhood was filled with Italian-Americans—the Lobaccaros, the Milanas, the Adamos, the Brunos. Ours was a big, extended Italian family. My mom only had one sister, Connie, who stayed in Brooklyn, but my dad had five sisters, who all settled somewhere in Queens or Long Island. My uncle Frank and aunt Ange lived right across the street from us, and my uncle Tony and aunt Marie were around the corner. My uncle Romeo and aunt Ann were nearby in College Point, Queens. I had aunts and uncles and cousins around all the time.

Franklin Square was blue-collar. Lots of guys worked for the city. A few guys you didn't quite know what they did, but they drove a big Lincoln and would stick wads of money in your pocket at weddings. Some kids went away to college, but most of them enrolled at the local school while living at home. A lot of guys became policemen. Your dad was a cop, so you became a cop. That was the mentality people had. My cousin Peter is crazy smart, and when he got into Princeton my aunt Sally burst into tears and cried and wailed and begged him not to go because she didn't want him leaving the family to go to school . . . in New Jersey.

My world was very small. People didn't think about leaving Long Island, let alone going to space. My buddy Q's dad was a pharmacist and his mom was a schoolteacher; he was one of the few friends I had whose parents had been to college. My parents always encouraged me to do whatever I wanted, but—being a fire inspector and a seamstress—there wasn't much they could do to help me become an astronaut.

I wanted to go to the Hayden Planetarium at the American Museum of Natural History more than anything; it was a big deal when my mom and dad finally took me. I brought home pictures of the planets and books on astronomy. But that was my only exposure to the world of space. How you got to join NASA or what college you should go to in order to get there—I didn't know anyone who could answer those questions. There was no science club at school where we could build and launch rockets. None of my friends were into space; it was something I did on my own. I had my spaceman costume, my Astronaut Snoopy, and my library books, and that was it. I didn't even know anyone who had a telescope.

Even if I had, I wasn't the most obvious candidate to become a guy who gets launched into orbit. I'd never been on an airplane. Part of the reason I idolized astronauts was because they were everything I wasn't. They were fearless adventurers, and I was an awkward kid. By the time I hit junior high, my vision was bad. I wore glasses. I was so tall and so skinny I could have been my own science experiment: If you wanted to know where the bones are on the human body, all I had to do was take off my shirt and I could show you.

Astronauts coming back from space had to splash down in the water, and I hated the water. I didn't know how to swim that well. Because there was no fat on my body, whenever my parents took us to the beach or to the local pool, it was like getting into an ice bath. I was scared of heights, too. Still am. Standing on a balcony four or five stories up and looking over? No, thank you. I didn't like roller coasters, either. They're scary. Hanging upside down? It makes you sick. Who wants to do that? I wasn't any kind of thrill seeker at all.

I had this fantasy about going to the moon, but that's all it was: a fantasy. The whole idea of actually joining NASA and going to space was so far-fetched and so far removed from my life that it was hard for me to stay interested in it. None of my friends seemed interested, and I wanted to be hanging out with my friends. What they cared about was baseball. Back then there were two leagues you could join in Franklin Square: Little League, which cost $15, or the Police Boys Club, which cost $5. In Little League you got the nice jersey and played on the nice field. In the Police Boys Club you got a T-shirt, and you played on the field that was mostly weeds and dirt. The kids with money played Little League. Me and my friends played in the Police Boys Club.

Soon I was as deep into baseball as I'd ever been into space. I was always throwing a ball. If I didn't have anyone to play with, I'd throw it against the stoop for hours, pretending I was pitching in a big game. The moon was 238,900 miles from Earth, but Shea Stadium was only twenty minutes away down the Long Island Expressway. My dad and I went to a lot of games, usually with my uncle Romeo and my cousin Paul.

As I got older the whole astronaut fantasy went away. It burned bright and burned out, like many childhood dreams do. It was the same for the rest of the country. The Apollo program stopped in 1972. By then the thrill of the space race was over. America had won and people moved on. I did, too. The astronomy books went back to the library, my Astronaut Snoopy went on a shelf, and by fifth grade I'd mostly forgotten about space. For a kid like me, being who I was, coming from where I came from, saying "I want to grow up to be an astronaut" was like saying "I want to grow up to be Spider-Man."

How the heck do you do that?

2

MOST ALL-AROUND

In my senior year of high school I applied to the engineering school at Columbia University on the Upper West Side of Manhattan. That November I went there for an interview, and the minute I arrived, I felt like I understood what college was. Before that, to me, college was a thing people did to get a job; but walking on that campus on this beautiful fall day, seeing the students hustling around and going to class, standing between Low Library and Butler Library, which look like something out of ancient Rome, I had a revelation: This is where people learn. This is where you become someone. I'd never had that feeling before.

To be honest, I didn't know if I belonged at an Ivy League school like Columbia. Growing up, I was never the smartest kid in class. I was a good student, but I wasn't exactly a genius. I liked science and math. I played sports, but I was just okay. My greatest talent was for people. I wasn't one of the popular kids, but I got along with everybody. I was enough of an athlete to hang with the jocks and the cheerleaders. I played the trumpet in band, which got me in with the band kids. I was in advanced math, so I could eat lunch with the smart kids. I moved in a lot of circles, had pockets of friends in every group, and learned how to mix with anyone.

I've always been curious about other people's lives. I find them interesting. I meet people and I want to know their story, what makes them tick. And the fact that I could hang out with different kinds of people helped make me a well-rounded person. I wasn't the smartest or the most athletic. I was the most all-around. If anything, it was my talent for getting along that made me stand out in school. At a parent-teacher conference, Mr. Stern, my eleventh-grade social studies teacher, said to my mom and dad, "Mike should think about applying to an Ivy League school. I think he'd do well." My parents came home and told me what he'd said, and that was the first time it had ever occurred to me to think of going to Columbia.

I submitted my application, but I didn't think I'd get in. I'd applied to a couple of schools on Long Island and I was convinced that's where I'd end up. Then one day a few months later I was at home, sitting on the toilet, when my mom came and knocked on the door. "You got a letter from Columbia." She slid it under the door and I opened it up. When I read the word "Congratulations," I started screaming. I was going to Columbia as a freshman that fall.

Columbia opened up a new world for me. I was only a few miles from home, but it was as if I'd landed on Mars. There were students from other countries, from fancy prep schools. Even Barack Obama was on campus at the same time I was. (Unfortunately, I never got to meet him, as I imagine being best pals with the future president of the United States would have some advantages.) As exciting as this new world was, I didn't fully embrace it right away. Whatever potential Mr. Stern saw in me, I hadn't found it in myself. I didn't take advantage of everything I was being offered. The summer after my freshman year, when the other kids took internships or went to study abroad, all I wanted to do was go back home. I moved back with my parents and got a job as a laborer in Rath Park, which is the park in Franklin Square where we used to play ball. I picked up trash, mowed the grass, cleaned toilets.

Change doesn't come easy for me. I liked my hometown. I was comfortable there. It was hard for me to leave, and I think deep down I knew that doing well at Columbia would mean leaving. Being an A student at an Ivy League school would put me on a new path. I was afraid it was going to pull me away from my hometown and my friends whether I wanted it to or not.

For my major I'd picked industrial engineering, which is about optimizing systems and organizations. It's called the most human- istic of the engineering disciplines. I liked the fact that it had a mix of hard science and traditional engineering courses along with courses in economics and business. Industrial engineering also in- cluded something I found interesting: human factors, which fo- cuses on designing machines and systems with human operators in mind.

I did fine my freshman and sophomore years, but junior year the course work got harder and I hit a wall. It was bad. I got clobbered in Circuits and Systems, an electrical engineering course. The mid- term counted for a quarter of the final grade. The average for the class was somewhere in the eighties. I got an eleven. On another midterm I got a fifteen.

Failing those tests turned out to be a good thing. It was a wake-up call. I was forced to decide what I wanted. At first I hon- estly thought about giving up. I contemplated changing my major out of engineering to something less technical; I didn't think I could hack it. But then I started thinking about my father, working as hard as he did for the city to give me the chance to go to college. We couldn't afford for my dad to take the Long Island Rail Road in to work. It was too expensive. He took the bus and the subway in to the city every day, over an hour each way, which was miserable. Up before dawn, never getting home before dark. I didn't want him to have done it all these years just for me to give up. I also didn't want to end up doing that myself.

So I went back to class and buckled down, studying as hard as I could. I went to a completely different level. My TAs were great. They spent extra time with me going over everything. My friends let me share their notes and spent hours helping me out. On the second midterm for Circuits and Systems, I got an 80. On the third, I got a 100. I went from the lowest score in the class to the highest. That semester I turned everything around. I made the Dean's List for the first time.

I also started thinking about the space program again. For most of the 1970s, like me, America had lost interest. We'd put up our first space station, Skylab, but it never captured people's imagination like Apollo had. Now we were entering the shuttle era. NASA had started doing drop-test flights out at Edwards Air Force Base in 1977 and had been flying operational space missions for over a year. The shuttle itself was very cool. It wasn't some tin can being launched into orbit. It was a real, honest-to-God spaceship. It took off like a rocket with up to seven astronauts, orbited the Earth, and gave the astronauts a place to live and work for a few weeks. We would use it to launch huge pieces of equipment and satellites, conduct experiments, and bring the results back to Earth. It was a sign that the space program was taking its next giant leap.

To me, the most exciting development was the new astronaut corps, which was full of interesting people: thirty-five in the first class, as opposed to seven for Mercury. And astronauts weren't just military test pilots anymore. The candidate pool had grown far wider. It included women and people of color. There were new faces at NASA, a new story to tell. Sally Ride had been chosen to be the first American woman in space, and her flight was scheduled to happen that summer. The anticipation surrounding her flight was huge. She was on every magazine cover and every news show. We were entering a new space age, and America was excited again.

Even with these new people joining the astronaut corps, I still wasn't thinking I'd ever be an astronaut; that dream was dead and gone. But one of the instructors in my mechanical engineering lab, Professor Kline, started talking to us about how private aerospace contractors—companies like Lockheed, Grumman, McDonnell Douglas, Martin Marietta—were getting big government contracts to work on the space shuttle systems. I thought maybe I could work as an engineer for those contractors, supporting the astronauts as a part of the bigger dream.

Back when the *Apollo 11* guys landed on the moon, I believed what they were doing was the most important work of our time: exploring space, pushing the boundaries of human knowledge about the universe. I never stopped believing that. If there's one thing I got from my father and his job, it was understanding the importance of public service. He instilled that in me. Schlepping an hour into the city on a bus to go around to gas stations and warehouses and inspect fire extinguishers and safety exits may seem like a menial job, but my father took great pride in it. He knew that people's lives depended on him doing his job well. He knew that firefighters were counting on him to make their jobs safer by preventing fires before they started.

The camaraderie that firefighters have, that brotherhood that forms among them—my father was a part of that, and it came from having a shared sense of purpose. He told me that whatever you do in life, it can't just be about making money. It's important that you work to make the world a better place, that you help improve the lives of the people around you. That's what I thought about when Professor Kline started telling us about the opportunities coming up with the space program. Everybody knows that firefighters are heroes, but they rely on guys like my dad to help them do their job. I thought maybe I could do the same thing with NASA.

Near the end of my junior year, I submitted applications to every engineering company on Long Island, and I got a summer job at Sperry, located in Lake Success, not far from Franklin Square. Sperry made everything from military hardware to office type-writers to electric razors. It was perfect. I could live at home, save some money, and still get some actual, hands-on experience.

That summer, everything was moving in the right direction. I was feeling good about myself. All I was missing was a girlfriend. I hadn't had much luck in that department, but that had a lot to do with my not having my life together. Now I felt like I was finally in a position where I was ready to meet someone.

My friend Mike Lobaccaro was working as a lifeguard at a pool in nearby New Hyde Park. I went over to pick him up one after-noon, and as I was waiting for him by the indoor pool, this girl, another lifeguard, was in the water giving swimming lessons to a bunch of kids. I thought she was really cute. Mike told me her name was Carola Pardo. A couple weeks later, she showed up at Mike's twenty-first birthday party at a bar in Mineola. At some point during the night Carola and I started talking. She was the same age as I was, about to start her senior year at Fordham, in the Bronx, where she was studying to be a physical therapist. We were both Sicilian-American, but she was of a more recent vintage. My grandparents had come over at the turn of the century. Her parents had come over in the late 1950s. I remember we were standing by the Ms. Pac-Man machine. She had these red cap shoes on, a jean skirt, and a rainbow-striped sweater with short sleeves. We started talking probably around 9:30 and that was it. By a quarter to mid-night we both looked up and the place was empty. The bartender was falling asleep behind the bar, waiting for us to finish. A few weeks later we were officially a couple.

Around the same time, at Sperry I met another important per-son in my life: Jim McDonald, an engineer who worked a few desks over in the bullpen. I walked up to his desk to ask a work question and ended up hanging out there for over an hour; I don't think we even got to the work question. Jim had a big crop of straight hair parted to the side and a friendly, whimsical smile. We hit it off immediately, and he became something of a mentor. More than a mentor, really—a guardian angel. He started watching over me, checking in on me when I was back at school, always making sure I was on the right path.

Sperry was my first experience with being an adult. My official job title that summer was "engineering aide," what you'd call an in-ternship now, except I actually got paid. My team designed inven-tory systems and conveyor belts for military warehouses. It wasn't exactly exciting. I had to dress like a grown-up, put on a white shirt and tie, and drive to work. Every morning me and a bunch of guys in white shirts and ties would file into this big building, work at a desk, go home, and come back the next day and do the same thing. Over and over and over again. Lunch was the high point of the day.

Some people want that. They like the routine, the safety of a paycheck every two weeks. But it wasn't for me. It didn't give me the sense of purpose I was looking for. As I got to know Jim Mc-Donald better, I realized he didn't care for it, either. Jim was not a typical engineer. He was very philosophical, more interested in the person he was talking to than the work that needed talking about. He'd just gotten married and had his first kid. At the time I thought he was so old, way older than me: He was probably thirty-six or something, but when you're in college, that might as well be a hundred. One day he said to me, "You're not enjoying this, are you?"

"No, not really," I said.

"Look, Mike. You don't want to end up here. I've been here ten years. I've got a mortgage, a kid. It's too late for me. I love my family. I do things on the side to keep life interesting, but you've still got a chance. You need to find something that you're passionate about."

He pointed to a guy a few desks away who was sitting there, bored, reading some science fiction novel. "You see that guy?" Jim said. "That guy just got his master's from Cornell. Do you know how smart he is? And look at what he's doing. You don't want that to happen to you."

For some reason, like Mr. Stern, Jim McDonald saw something in me, some kind of potential. I hadn't thought seriously about being an astronaut since I was seven years old. At that point the dream was dead. Jim opened the door for it to come back to life. That whole summer, right up to the last day I left and headed back to Columbia, he kept pushing me. "Go and do something different," he'd say. "Go to grad school. Find something meaningful. Find something important. Whatever you do, don't come back here."

3

WHO YOU GONNA GET?

When my final semester at Columbia came around, my boyhood astronaut dream was still dormant. Then, on one Saturday evening in January 1984, my whole world changed. I was home in Franklin Square for the weekend, and Carola and I decided to go to the movies to see *The Right Stuff.* From the balcony of a theater in Floral Park we watched the story of the original Mercury Seven astronauts: Alan Shepard, the first American in space. John Glenn, the first American to orbit the Earth. These fearless test pilots pushing the envelope, risking their lives to help America win the space race against the Soviets.

It was *awesome.* These astronauts weren't just doing this big, important thing for their country, they were also having a blast. They were flying fighter jets through the clouds, racing convertibles across the California desert, wearing leather jackets, smiling behind their cool aviator sunglasses. They were risking their lives every single day on the job. They were the baddest guys I'd ever seen. I didn't take my eyes off the screen for one second.

One scene in particular blew me away. John Glenn is all set to be the first American to orbit the Earth, but his launch is aborted. Vice President Lyndon Johnson is waiting outside Glenn's house, demanding to bring in TV crews to talk to Glenn's wife, Annie, on national television. But Annie has a stuttering problem; she doesn't want to be on TV. So John gets on the phone with her and basically tells her it's okay if she wants to tell the vice president of the United States to get lost. Some suit from NASA jumps down Glenn's throat, telling him he can't blow off the vice president like that. Glenn won't back down. So the NASA guy threatens to yank him out of the flight rotation and replace him if he won't toe the line. Then the other Mercury guys step up and get in the guy's face, and Deke Slayton says, "Oh, yeah? Who you gonna get?"

Finally Alan Shepard tells the suit, "Step aside, pal." They've got Glenn's back.

That moment, to me, summed it up. That's how you treat your buddies. You stand up for each other. You stand up for what's right. I saw that and I said, "I want to be one of those guys." I wanted to go to space, but more than that, I wanted to be part of that team, to have that camaraderie, that shared sense of purpose that comes from doing something big and important. That's what was really cool about that movie—that and the view from space. When John Glenn is in his capsule looking down on Earth, the expression of wonder on his face, that floored me. The second I walked out of the theater, I knew: I wanted the whole enchilada. I wanted to be an astronaut.

My next thought was: *How the heck am I going to make that happen?* I was on the verge of graduating from college, but because my astronaut dream had been dormant for so long, I hadn't mapped out my education with space travel in mind. Columbia is a great school and it gave me a great education and a solid foundation, but back then it wasn't a traditional pipeline to the astronaut program. I was also an industrial engineering major, which didn't seem like the right major for becoming an astronaut at all; I felt like I should have done mechanical engineering or aerospace engineering.

I had made one good decision. Back when I was working at Sperry, Jim McDonald had told me about the Program in Science, Technology, and Society at MIT. It concentrates on how scientific progress affects and interacts with other aspects of life, like public policy and how people live; it's a degree for people who want to contribute to more than just the technical side of things. That sounded interesting. I sent off my application to this program and I waited. I didn't do it with any thought of being an astronaut; at that point I hadn't seen *The Right Stuff* yet, and being an astronaut was still the last thing on my mind. But if you do want to be a part of the space program, MIT is one of the best schools to attend. By total coincidence, I'd taken at least one step down the right road.

While I waited to hear about grad school, I started looking for work. I still wanted to follow my father's example and work in public service. One week IBM came to campus to recruit students from the engineering school, and I talked to one of the interviewers about their public sector office, which worked with nonprofit institutions to set up and service their computer systems. I felt like that might be interesting and rewarding at the same time.

IBM asked me to come in for an interview, but the day before I was scheduled to go I got a letter from MIT. When I opened it (I wasn't on the toilet this time) I was shocked: They actually let me in.

The next morning I went for the interview with IBM and it went well; the job was mine if I wanted it. I told the interviewer about the MIT offer, and he said IBM employees took leave to go to school all the time. I could defer school, work for IBM for a couple of years, then take a leave to go to MIT and they would save my job for me. I just had to figure out what worked better: MIT now and work later, or work now and MIT later.

The first thing to do was to visit MIT. I called my dad, who took a day off from work, and we drove up to Cambridge to meet with the director of the Program in Science, Technology, and Society. He was a weird-looking academic type with crazy hair. We started talking and he seemed confused as to how I had ended up in his office. Apparently I hadn't read closely enough when I was researching the program. It wasn't a part of the engineering school. It was a political science degree. We were in the political science department. My father looked at me and said, "What are you doing in the political science department?"

I'd applied to the wrong grad school.

I didn't even know MIT *had* a political science department. Science, Technology, and Society, it turned out, was a program for people who wanted to write papers on how science is affecting society. The similar-sounding but totally different program in the engineering school was called Technology and Policy. It was also about how technology impacts society, but it was for engineers and scientists who actually want to design and build things. MIT allowed me to resubmit my application to the engineering school, and luckily I got in there, too.

Even with that sorted out, I wasn't sure what to do. I'd only applied to MIT because Jim McDonald had told me it might be a good idea. I never expected to actually get in. I had no idea what I would study there, what my research would be. I hadn't thought about any of that. I didn't have any way to pay for it, either. I didn't have any scholarship or fellowship, and I knew my parents couldn't foot the bill. IBM had a great training program, and working there would keep me in New York near home. Plus I knew I could take the job and earn money to go to grad school later. MIT felt like this huge unknown, a stretch, a risk. IBM felt like the safe choice.

I made the safe choice.

After graduation, I moved back in with my parents and commuted every morning on the train to IBM's building at Fifty-Seventh and Madison in Manhattan. The job seemed to suit my personality well. I was the technical side of the sales team assigned to the Port Authority account. Once or twice a week I'd go down to the World Trade Center, work with their information technology guys, take people to lunch. The sales team was also responsible for the entertainment at IBM's monthly branch meetings. We'd put on little skits about the slow elevators and the bad food in the cafeteria. I was making a decent salary and people were treating me like a grown-up. And IBM was a great company: They took care of people. But something was missing. I didn't have that sense of purpose I was looking for.

Then, on the Fourth of July 1985, *The Right Stuff* came out on HBO. My parents didn't have HBO, but my friend Mike Q did. He let me make a VHS tape of the movie off his TV. Every night I'd come home on the train, pop the tape in the VCR, and watch it—and I mean literally *every* night. I'm not exaggerating. I'd stay up all the way to the end, watching Chuck Yeager push his Lockheed NF-104A up and up and up to the edge of space only to come crashing back down to Earth—and still walk out alive, chewing a stick of Beemans gum. Then the next morning I'd wake up, put on another white shirt, get on the train, and go back and sit at my desk.

Going to IBM wasn't a mistake. It was something I needed to do in order to realize it wasn't what I wanted to do. Carola and I were getting serious, and I figured we were going to get married. If I stayed where I was, we'd end up living in New York somewhere, taking the train every day, and that would be it. I was only twenty-two years old, still living at home with my parents, and I could already see my whole life being over, mapped out and done.

That summer, the last week of July, I decided to drop in and see Jim McDonald, my old mentor from Sperry, on the way to a Mets game. We went out and played catch in the street for a few minutes. We were tossing the ball back and forth, and he asked, "What's going on with you?" I told him about IBM, making the sales calls at Port Authority, doing skits for the branch meetings. He stood there and gave me this look.

"What's the matter with you?" he said. "Imagine the conversations we'd be having right now if you'd decided to go to graduate school. You'd be telling me about hearing lectures from Nobel Prize winners. You'd be telling me about the exciting new research you're working on. MIT is the opportunity of a lifetime. Instead you're telling me about what, doing skits in some office in Manhattan?" He whipped the ball at me and it landed in my glove with a pop. "You need to wake up," he said. "Don't blow this."

Talking to Jim, I realized part of my problem was that I didn't have anyone to talk to. He could give me pep talks, but I didn't know anybody who was involved in the space program. I didn't even know anyone who knew anyone who was involved in the space program. I figured I should go to grad school, but what should I study there? What did I need to learn?

Part of what I loved about *The Right Stuff* was the camaraderie. Getting to space isn't something you can do on your own, and I was on my own. I had lots of friends, but I didn't have any space friends. I needed space friends.

Out in Garden City, Long Island, not far from Franklin Square, they have the Cradle of Aviation Museum. It's built on Roosevelt Field, where Charles Lindbergh took off for his flight across the Atlantic Ocean. The weekend after I talked to Jim, they were putting on a fair to celebrate space flight. My mom clipped an article from *Newsday* about it and saved it for me. I decided to go and see if I might meet anybody.

One of the booths at the fair was for the Civil Air Patrol, the civilian auxiliary of the Air Force. There was this kid working there, this little Italian guy named Mario, all dressed up in a Civil Air Patrol uniform. We got to talking and it turned out he wanted to be an astronaut, too. Only he wasn't just dressed like a pilot. He *was* a pilot. He was only sixteen years old and he already had his private pilot's license. When I was his age I could barely drive a car, and he was already flying airplanes. I started pestering him with questions, and he had his whole plan laid out: his application to the Air Force Academy, what kind of jets he wanted to train on, the whole nine. The twenty-two-year-old Ivy League graduate was desperately hoping to learn something, anything, from a sixteen-year-old kid. I felt like an idiot.

I decided to write a letter to NASA. I had no idea who to send it to, so I sent it to the top guy, NASA administrator James Beggs. He didn't write back, but I did get a reply from a man named Frank Coy, Beggs's executive officer. I guess he's the guy who got the junk mail, like my letter. For some reason he wrote back and said to give him a call. I did. He told me about different jobs at NASA and the different aerospace contractors. The upshot of the conversation was that no matter what I did, if I wanted to have any chance of being an astronaut, I had to go to graduate school.

Every morning I'd pick up the *New York Times* at the Long Island Rail Road station on the way to work and read about the latest developments at NASA. The shuttle program was going at full steam by that point. Flights were going up every six weeks. They were deploying satellites, conducting Spacelab missions, launching secret payloads for the defense department. My dreams were playing out on the front page of the morning paper every day, and there I was, on the train, reading about it instead of doing it.

Even though MIT had accepted me, I still had a hard time seeing myself there. I think my single biggest problem was that part of me believed I was supposed to be on that train. Even when I was at Columbia, I was a kid who thought I'd live on Long Island for the rest of my life. I was going to hang out with the same guys, watch every single Mets game, and be content with that life. In some ways, part of me still is that guy. But there was also the other part of me, the kid who walked out on his front lawn and looked up at the moon and dreamed of going there. I realized much later in life that the reason this decision between MIT and IBM was so agonizing was because it wasn't really about choosing a career; it was about deciding who I was, which part of myself I wanted to be, and that's the hardest decision any of us has to make.

I took a day off from work and drove up to visit MIT. I talked to some students and professors who were designing equipment and experiments to be used on the space shuttle. They were researching how the human body functioned in space, how to control robots on other planets. Once I saw that, I knew MIT was where I had to be. How I would pay for it, where Carola and I would end up, I had no idea. But I knew in my heart I had to at least give it my best shot.

If I had any remaining doubts about grad school, they disappeared the morning of January 28, 1986. I was at my desk and a coworker came by and said, "Mike, did you hear? The shuttle exploded." Somebody turned on a television in the reception area and we rushed in to watch. It was on every channel, playing over and over again: The space shuttle *Challenger* had exploded in this giant ball of flames, the debris flying off in a Y-shaped trail of smoke. The O-ring in the right solid rocket booster had failed, leaking burning-hot gas that had caused the explosion. Seven crew members were on board: five astronauts, Francis Scobee, Michael Smith, Ron McNair, Ellison Onizuka, and Judy Resnik; Greg Jarvis, a payload specialist; and Christa McAuliffe, a schoolteacher chosen to be the first civilian in space.

All shuttle flights were suspended, and would be for the next two and a half years. I was going off to grad school to try to be an astronaut, and the whole space program was more or less on hold. But that didn't matter to me. It was strange, but after two years of wavering, once the accident happened, I never second-guessed myself. When the *Challenger* exploded the world stood still. The president came on TV. Everybody paid attention. It reminded me how important the space program is. The world honored the seven people on board the shuttle that day because what they were doing—the sacrifice they made—was so important. They had the Right Stuff.

I knew right then that I wanted to be a part of something that meaningful. I wanted to have something I was so passionate about that I'd be willing to risk everything for it. I wanted to know that if I ever got killed, I got killed doing something worthwhile. The kid who looked up at the moon and wasn't afraid to dream—I decided that part of me deserved a chance. I sat there in that reception area, watching the crash footage play over and over again on the television, and that was when it hit home for me: You only have one life. You have to spend it doing something that matters.

part

2

Maybe You're Not
Cut Out for This

4

THE SMART-KID OLYMPICS

MIT may be the most intimidating place on Earth. I felt like I was out of my league the day I showed up. The grad students you find at MIT are the ones who kick themselves because they "only" got a 790 math on their GRE. I did not have a 790 math on my GRE. Nowhere close. And it's not just the smartest American kids. MIT draws brilliant people from all over the world. I made several friends from Algeria. Back then, every year, the Algerian government would take their top two or three engineering students and pay their way to go to MIT so they could come back and become professors at their universities. Algeria may not be the biggest country in the world, but you take the top two engineering students in all of Algeria and those are going to be some pretty smart people. It's the same with nearly every other country on Earth. It's the best of the best from Thailand, from Brazil, from Poland. It's the smart-kid Olympics.

Then there was me: the guy from Long Island who filled out his application wrong.

What helped me to find my place at MIT was the same thing that helped push me toward Columbia: Somebody saw something in me. Once I'd decided on grad school, I started going to the New York Public Library on my IBM lunch breaks to learn what MIT was doing with the space program. I started reading about Professor Tom Sheridan, who was doing important work with human factors and robotics.

Human factors is fascinating; it's what drew me into studying industrial engineering at Columbia. Anytime you get in your car and you can work the brakes and the steering wheel and read the speedometer and not drive off the road in confusion, that's because an engineer who understands human factors designed it for you. There's the engineering side of it, which is designing and building the machines, but there's also the human operator side of it, which is understanding how the brain responds to different stimuli and how to account for that in your designs.

Tom Sheridan was a professor of mechanical engineering and a professor of applied psychology, and he was like a rock star in the world of human factors. He was also doing cutting-edge work with the space program, designing control systems for telerobotics— how an operator on Earth can accurately manipulate machines and systems on satellites or the space station or even other planets. He also seemed like a cool guy, a good man.

I made an appointment to go up to Cambridge and visit him. His office was cluttered with piles of books and papers everywhere, a bicycle stashed in the corner. He was a bit of an absentminded-professor type, sort of gray-haired and disheveled. But he was warm and friendly and down to earth, a thoughtful, caring person among the crazy hard-driving personalities at MIT; he kept a big poster on his wall with a photo of the Earth that said "Love Your Mother." When we talked he said he liked the fact that I was an industrial engineer, because it had given me some practical experience with human factors, something that most of the *Good Will Hunting* geniuses on campus didn't necessarily have. When I walked into Sheridan's office he was like *Here's a kid I can work with.* He told me if I came to MIT, he'd love to have me in his lab.

Life is funny. I'd applied to the wrong graduate program, but that eventually led me to the right grad program. I'd taken what I thought was the wrong undergraduate major, and that was the thing that set me apart and allowed me to find my niche. I don't know if there are any lessons to take from that except to realize that the things you think are mistakes may turn out not to be mistakes. I realized that wherever you are, if you make the most of what you've got, you can find a way to keep moving forward. I packed up my office at IBM on July 4, 1986, and I moved to Cambridge, Massachusetts: the first stop, I hoped, on the road to space.

For the next six years I had my head buried in books. That first semester was stressful, but incredible. I got a half-time teaching assistanceship that covered half my tuition and gave me a small stipend. I was taking three classes: Tom Sheridan's class, which was more of a project class; a technology and policy pro-seminar; and an economics class. It was tough, but I did well: two As and a B. Then, the second semester, I got destroyed. I took my first aerospace class, a satellite engineering class taught by Professor Walter Hollister, a short guy with a gigantic mustache who had flown fighter planes in Vietnam and became a PhD in aeronautics and astronautics. (A fighter pilot *and* a professor—how much cooler could you be?) The first exam was brutal. When the test came back I got a 35, the lowest grade in the class. Mark Stephenson was a buddy of mine, a really sharp guy from West Point. He walked up to me after the test and he said, "How'd you do?"

I said, "I'm toast."

I told him about my 35. He had done better than me, but not by a whole lot. We ended up talking to this other guy, Wasif, who got a 38. Wasif was Indian but he'd grown up in Scotland, so he had this thick Scottish accent and he could drink like a Scotsman, too. Somebody said, "Let's get a drink." That seemed like a great idea, and we headed over to the Thirsty Ear, the MIT grad school bar. We sat around for hours, commiserating about our grades and how tough it was at MIT. It was a relief to find out I wasn't the only one struggling. Mark and Wasif and I ended up forming a study group to help each other out. We'd stay up late, ordering pizza, doing problem sets together, and going over the notes night after night after night.

It got tougher my second year. I decided to try to get two master's degrees at the same time; on top of the technology and policy degree, I was going for a master's in straight mechanical engineering, which I knew I'd need for the work I wanted to do with NASA and the astronaut program. Fall semester, I had four incredibly tough classes. I took dynamics from Stephen Crandall, who wrote the book on dynamics. I took mathematical principles for engineers from Gilbert Strang, who wrote the book on mathematical principles for engineers. The education I got was incredible, and not just the one in the classroom. I remember Sheridan telling me once, "If you can learn to live with indignities in life, you can go far." And he's right. You can learn a lot by getting knocked down, and I got knocked down over and over again. And every time I got up and kept going. I know there were students in my class who were smarter than me, but I don't know if there was anybody who worked harder than me.

. . .

Twelve men have walked on the moon. Four of them went to MIT. If you want to be an astronaut, going to a school like MIT is like going to Hollywood if you want to be a movie star: It's where you need to be. Your dream might not come true, but you're there and you're going for it and nearly everyone around you is going for it, too. You learn from them. You see what works and what doesn't. Once I got to MIT, my wanting to be an astronaut didn't seem so nutty anymore. It wasn't like wanting to be Spider-Man at all. It was the longest of long shots, but it was still something real people did—it was possible.

It seemed like every other day at MIT you were running into former and future astronauts: Franklin Chang-Díaz, an MIT grad who holds the record (in a tie with Jerry Ross) for flying the most space shuttle missions. Byron Lichtenberg, an MIT grad student who'd actually flown as the first payload specialist on STS-9. The first day of school I met Dava Newman, one of my fellow classmates; Dava never applied to the astronaut corps, but she went on to become a full MIT professor and serve as deputy administrator of NASA, the number two person in the whole space program. There were two students in my lab who wound up being astronauts with me, Dan Tani and Nick Patrick—and my lab only had about ten people total at any given time.

Those were the kinds of people you'd run into walking up and down the halls. MIT's Human-Machine Systems Lab was also where some of the world's most important academic work for robotics and the space program was taking place. These were experiments looking forty, fifty years down the line in terms of the future of space flight. Some of the work was out there. People at MIT had wacky ideas, but they were good wacky ideas, grounded in real science, not science fiction. Students in my lab were working on the human-operated control of deep-sea submersibles, like the robots they sent down to find the *Titanic*. Some were doing work on controlling robots on other planets, pioneering the technology that would allow us to land a rover on Mars in 2012, nearly twenty-five years later. And now I was in the mix with these people. Every day my dream felt more and more real.

At the end of my first year I went back to New York for a few weeks. Carola and I got engaged, but we decided not to get married for a couple of years, and she stayed in New York. I wanted to get some real experience with the space program that summer, so I wrote to Frank Coy again, inquiring about a summer job at NASA headquarters in DC. Most of what goes on at NASA headquarters is bureaucratic, administrative work, not much research or actual building of hardware. But I wanted to get a general introduction to the space program, and headquarters seemed the best place to do that. I applied for and got a position cataloging the human factors work NASA was doing and compiling the information in a report for the NASA administrator.

I spent that summer in DC drinking everything in. NASA headquarters is close to the National Air and Space Museum. I'd go on my lunch breaks to walk around. They had a brand-new IMAX theater that was showing *The Dream Is Alive*, the first IMAX film shot aboard the space shuttle. I think I went to see that movie at least fifteen times. But the most valuable thing I did was meet people. I met *everyone*. I'm sure I was the youngest person in the building by a good fifteen years, so I stood out to begin with, and I made a point of shaking hands with and talking to every single person I could. I met the then-current NASA administrator, James Fletcher. I met J. R. Thompson, the director of the Marshall Space Flight Center in Huntsville. Many former astronauts wind up working at

NASA headquarters, too. I wrote to them and asked if they would meet with me to talk about becoming an astronaut. Bob Crippen, the first pilot of the space shuttle, encouraged me to write to George Abbey, the head of flight crew operations in Houston, to get my name in front of him as well.

People could see my excitement, and with me being a Columbia graduate and an MIT student, they knew I had potential. My dream didn't feel so crazy anymore, but the reality was sometimes mind-blowing. One afternoon I walked into the cafeteria and Michael Collins was there having lunch; he was in town for a meeting to discuss a possible mission to Mars, where I'd seen him earlier that day. Here was one of the *Apollo 11* astronauts, a guy I'd read about and idolized for years, and he was sitting at a table eating by himself. I was beyond intimidated, but I knew if I didn't approach him I would regret it for the rest of my life. I took a deep breath and walked over and asked if I could join him. "Sure," he said, "have a seat." He asked me who I was, what I did. I told him I was an engineering student at MIT, that I wanted to be an astronaut. He was friendly, chatted with me awhile, asked what I was doing in school. I asked him if he had any advice. He told me if I was serious I should look at working in Houston or Huntsville instead of Washington, which turned out to be useful information.

The more I talked to people like Bob Crippen and Mike Collins, the more I realized that they were once the same guy I was: a young person with an impossible dream. Two years earlier I'd been clueless, lost, watching a grainy VHS tape of *The Right Stuff*. Now, here I was, having lunch with one of my childhood heroes, and we were just having a conversation. He didn't talk to me like I was some delusional idiot who was wasting his time. He talked to me like I was someone who was supposed to be there in that cafeteria. That alone was worth the whole summer in DC.

In the spring of 1988, I graduated from MIT with two master's degrees: a master's of science in mechanical engineering and a master's of science in technology and policy. Now it was time for my next great challenge: driving to Alabama. I'd more or less begged my way into a fellowship from the NASA Graduate Student Researchers Program at the Marshall Space Flight Center in Huntsville. At the time I was driving this 1976 Ford Grenada—the Ford Disaster, I called it. This thing could barely make it around the block, let alone all the way to the Deep South. It had been in a flood, and the floor was rotted out; my father and I had to nail roofing shingles to the underside of the car to cover the holes. I loaded it up and prayed I didn't break down in the backwoods of Appalachia along the way.

I made it—barely. I found a cheap place to live and went to work in the Human Systems Integration Branch. My summer in Huntsville was pure fun. I assisted with some very cool robotics work. I was on the softball team. I met a lot of people my age, and we took road trips to Atlanta and Tennessee. In my free time I started building out my résumé with the kinds of things people told me I needed to be a good candidate for the astronaut program. I got scuba certified. That was challenging; I still wasn't a good swimmer. I started taking classes with a flight instructor to get my pilot's license, and I ended up soloing a plane by the end of the summer.

After two years at MIT and two stints with NASA, I felt like I was plugged into what was going on in the space program, not stumbling around in the dark. As I packed up my Ford Disaster and drove home from Alabama, there was only one cloud on the horizon, something I knew I had to go through but I'd been doing my best not to think about: my PhD.

5

FORCE FEEDBACK

The thing about becoming an astronaut is that no one can tell you how to become an astronaut. Even the people at NASA can't tell you how to become an astronaut, because the chances of actually becoming an astronaut are so small. Everyone is encouraging, but no one has any ironclad advice. It's not like being a lawyer, where people can tell you, "Go to law school, pass the bar exam, and you'll be a lawyer." It doesn't work that way. I could get five PhDs and still never be an astronaut, and not every astronaut has a PhD. What people told me was "You should get the PhD because you want the PhD, because having that degree will make you happy in and of itself, and the astronaut dream may happen or it may not."

Getting my master's from MIT had nearly killed me, and I knew going for a PhD would be worse. I'd been accepted to the doctoral program; back then, if you completed your master's, MIT would let you go for a PhD almost automatically. But I was on the fence about it. *Is this really how I want to spend the next four years of my life?*

Before going to Huntsville I'd sent out résumés to a bunch of aerospace contractors in Houston, thinking that might be the better route. Those contractors work hand in glove with NASA; they're up at the Johnson Space Center every day. I came home from class one day and found a message on my answering machine from Bob Overmyer at McDonnell Douglas. They were hiring and he wanted me to come down to Texas for an interview. Overmyer was a former astronaut himself, a backup from the Apollo years who later flew on some of the earliest shuttle flights. I called him back and we talked and he flew me down for an interview.

Bob spent the whole day with me. He was a great guy, a former Marine who looked the part with the graying crew cut and a big smile. We talked about becoming an astronaut and he told me I'd be a great candidate and he sold me on everything he and McDonnell Douglas were doing to help build the new International Space Station. He took me down to the NASA cafeteria for lunch and introduced me to a whole bunch of people. It was fantastic. But one thing I learned was that, in the aerospace industry, people with master's degrees in engineering are not hard to find. There are literally thousands of them crawling around Houston and Huntsville and every other pocket of the aerospace industry, all wanting to be astronauts and all waiting for their shot. It's hard to stand out. One of the guys who interviewed me for a job in Houston was a frustrated NASA astronaut applicant himself; he had a master's degree and he wasn't getting anywhere. At that point I didn't have anything on my résumé that would give me a leg up over anyone else. I was never in the military; I wasn't an ace test pilot; my academic credentials were solid but nothing out of the ordinary. Ultimately I decided that a PhD from MIT was not only my best chance of setting myself apart, it was also a remarkable opportunity I shouldn't pass up. Houston would have to wait.

Once I decided to go for my PhD, I had to decide what it would be. I knew I didn't want to do straight mechanical engineering. I wanted to be able to incorporate some of the work on human factors that I'd been doing with Tom Sheridan. What I was interested in was robotics and control systems: how robots interface with human operators and, more specifically, how they do it in space.

The title of the thesis I proposed was "Sensory Substitution for Force Feedback in Space Teleoperation." In English, what that describes is a problem with controlling a robot when you're dealing with a time delay. Whenever you manipulate an object with your hand, pulling a lever or twisting a knob, the amount of resistance it gives you is something you can feel. It's instantaneous, and you can react to it right away. The brain knows automatically how to read those signals and adjust to apply more force or less. But if you're manipulating an object remotely via a robot—for example, communicating with a rover on Mars—there's a time delay between the signals the robot is sending to you and the commands you're sending to it. You might push too hard or not hard enough based on wrong information about what's happening on the other end, and the object you're manipulating becomes unstable very quickly and you start knocking into things. It's a problem of force information, force feedback.

At that time, the way robot designers dealt with this issue was by having a display for human operators that showed the levels of force feedback visually, like a speedometer that ratchets up or down. The problem with that was that operators already needed to have their eyes on the display of the object they were manipulating; having another screen to look at didn't exactly make things any easier. My idea was that you could eliminate that screen and transmit the necessary force feedback through an operator's other senses: touch and hearing. A slight vibration on your skin or a sound in your ear would indicate the level and direction of force feedback, allowing you to respond accordingly.

Pretty cool, right?

To research my thesis, I'd have to work across three different disciplines—mechanical engineering, aerospace systems, and neuroscience, the latter to understand cognitive brain functions. Tom Sheridan agreed to be my advisor, and I approached three other professors to serve on my thesis committee: Richard Held, a cognitive brain guy; Dave Akin, an aerospace systems guy; and Nat Durlach, an electrical engineering guy who worked with electronics and human perception.

As a PhD candidate, once I started my research, I had to pass a qualifying exam. It included written and oral components that tested basic engineering concepts and a presentation of where I was with my research. It was a way of making sure I was on track to successfully complete my work and also to make sure my engineering knowledge was up to MIT PhD standards: "Quality control," Sheridan used to call it. Basically, they didn't want me hanging around, wasting their time, if I was eventually going to blow it.

Some PhD candidates take their qualifier about six months in. I figured I'd need at least a year to prepare, so I scheduled my exam for the summer after my first PhD year and got to work. My project was going to deal with some serious cognitive brain science— how the brain processes sensory input and information—and that's where I had the least amount of experience. I signed up for this one graduate-level neuroscience class and it only took two weeks to find out I was in the wrong place. This was a class that third-year medical students from Harvard would come over to take at MIT. We're talking the cream of the crop from one of the top medical schools in the country. The professor jumped in on day one assuming everyone in the room was a medical genius. Which everyone was, except me. I felt like I was in a foreign-language immersion class—that's how little I understood what was happening. The words he was speaking and writing on the board, I *think* it was Latin, but for all I could understand, it might as well have been that elf language from *The Lord of the Rings*. I knew I was way out of my depth. I dropped the class and took an undergraduate neuroscience course instead. That one was impossible, too, but at least it was in English.

That whole year was difficult. One of the only bright spots was that the shuttle had started flying again. In September, STS-26, the first flight since *Challenger*, launched from Kennedy Space Center for a successful four-day mission aboard the shuttle *Discovery*, which meant the space program was back in full swing and there would be more astronaut jobs in the future. I'd sent away for an application to be an astronaut and was working on it here and there. I didn't expect to get in; I knew most people got rejected on their first try. But I wanted to get my name in the hopper anyway.

Other than that, it was a gloomy time for me. Most of my good friends left after finishing their master's degrees. Carola and I were looking forward to getting married the next summer, but in the meantime we had to be content with seeing each other on weekends here and there. I was working alone, living alone. I'd wake up, trudge through the Boston winter to go sit in a carrel in the library, study all day, work in the lab, and go home to this empty dorm. Looming over me the whole time was this dreaded qualifying exam. I studied and studied and studied, but I had no idea what I was doing. I had no idea if I was studying the right material the right way. I'd never faced anything like this. I was just cramming information into my head and hoping for the best.

D-day finally came the week of June 22, 1989. That same week the Mets traded my favorite player, Roger McDowell, and future MVP Lenny Dykstra to the Phillies, one of the worst trades of all time. I took it as a bad omen. It was. The day of the oral exam and research presentation, I woke up terrified. I walked over to Professor Sheridan's office in the Mechanical Engineering Building, dreading what was to come. Sheridan and the rest of my committee were lined up in chairs around a coffee table and then there was me, alone, standing at a chalkboard in front of them with nothing but a piece of chalk in my hand to show my work.

I needed more than a piece of chalk to save me that day. An oral exam is like a firing squad. It's their job to tear you apart, challenge your assumptions, force you to defend your conclusions. If they find a weakness in your work, they'll hone in on it and take you down. Albert Einstein could be up there and they'd tear him to shreds, too.

I won't keep you in suspense: They destroyed me. It was a massacre. They were bombarding me with questions from every angle. Why this? What's your evidence for that? I was stumbling through my answers, losing my train of thought, trying to go back and start over. After a certain point, I was completely lost. They could have asked me "What's two plus two?" and I don't think I could have given them a clear answer.

There was one question about a system for a human operator to control a robot through a helmet that measured the movement of the human operator's head; as the person turned right or left, so would the robot. But the way the robot was built, the control system was inherently unstable. If the human turned his head a certain way, the robot would vibrate out of control. I was clueless to this, and I'd gone about modeling the system as if it were perfectly stable—I'd based a huge chunk of my answer on an assumption that was totally false.

Sheridan said, "Mike, what happens if the operator turns his head to the right?"

I said, "The robot turns to the right."

He said, "Is that *all* that will happen?"

I had no response. He kept pivoting his head to the right as he asked me this, to demonstrate what he was talking about. He was actually trying to help me, give me a hint. I had no idea what he was getting at or why he kept jerking his head to the side. I stared at him like a confused zoo animal. Then he explained the answer to me, and at that moment I knew it was over. I was toast.

I walked out of that room completely traumatized. I wandered around the campus in a daze. I'd blown my PhD. I'd blown everything. My NASA application was sitting at home on my desk, filled out and ready to submit. Now what was the point in that? Carola and I were getting married in a few weeks. She was already packing up and planning to move to Boston. We had a place picked out, this great one-bedroom by Harvard Square with a big bay window looking out on Massachusetts Avenue. Now that whole future had crumbled. What was I going to tell her?

I went back to Sheridan's office a couple of hours later. When he opened the door I could tell by his expression that the results weren't good. "I didn't pass, did I?"

He looked away, shook his head, and said, "No, Mike. You didn't pass."

Sheridan truly cared about his students; it wasn't easy for him to deliver that news, but he had to do it.

I had the option, if I wanted it, of coming back and taking the exam again in six months. But I'd failed so utterly and completely that he said to me, point-blank, "You should think about whether or not it's worth your time to do that." Sheridan had always been so supportive, a wonderful man. Without him I wouldn't have made it as far as I did. But here he was telling me, in so many words: *Maybe you're not cut out for this.* Hearing someone who had been so supportive be so brutally candid now was difficult to take. I sat there in his office, completely and totally demoralized. A year earlier I'd been walking on air: working at NASA, meeting my childhood heroes. Now everything had come crashing back to earth, and I had no idea how to get back up.

When I got home from blowing my qualifying exam, my completed NASA application was sitting there on my desk, like salt in my wound. My first thought was *Well, that's done.* Then I thought about it and decided to send it in anyway. I'd failed the test, but they were offering me another crack at it, and until I decided what to do I was still a PhD candidate at MIT, and many astronauts with only a master's degree do get picked. So I mailed it off without telling them, "Hey, and by the way, I just failed my qualifier!" I figured by the time anybody looked at the application I'd have another crack at the test, if I felt it was worth taking it again. I was all but sure it wasn't.

Then I went back to Huntsville. I owed them three more weeks of work for my graduate fellowship. The summer before I'd been excited, confident. Now I was totally down in the dumps. The three weeks I was there happened to coincide with a celebration for the twentieth anniversary of the moon landing—twenty years since I'd stood out on my lawn and looked up at the moon and dreamed about going there someday. They were having lectures and symposiums, and all the Apollo guys were coming through to give talks and be recognized. Neil Armstrong was there. Buzz Aldrin and Mike Collins and Pete Conrad were there. But instead of being inspired I was more like *Great, this will never be me. I'm never going to be one of these guys.*

One of the Apollo astronauts there that week was Charlie Duke, who flew with John Young on *Apollo 16*. He was also one of the four moonwalkers who came through MIT. I attended the talk that he gave, and after it was over he was signing autographs for people. I picked up an Apollo postcard and got in line to go up to the table where he was sitting. While he signed the postcard for me, he said, "So what do you do?"

I said, "I'm a research fellow here in Huntsville, and I'm a student at MIT."

He said, "MIT? Man, that place kicked my ass. I never thought I'd make it out of there, but somehow I did."

I stood there, thinking, *Wow. This guy walked on the moon, and even he barely made it out of MIT. And he might never have made it to the moon if he hadn't made it out of MIT.* I realized that Charlie Duke and the other Apollo astronauts, before they walked on the moon, had all walked a mile in my shoes. The journey to space wasn't easy, but if I gave up, it would be over.

In that moment I learned how much power astronauts have to inspire people. I walked away from that table and I knew: I had to go back. I had to try again. Maybe I wasn't a failure and an idiot. Maybe a PhD from MIT is something that's really hard and it knocks everyone down and forces them to get back up. And this wasn't only my PhD at stake; my whole space dream was on the line. I decided that, as bad as I went down in flames the first time, I had to turn it around and pass the test. That was the only happy ending to this story. The chances of my doing that successfully were pretty much zero, but if I didn't at least try I'd always look back and be disappointed with myself.

Carola and I got married the next month in New York, and we took a three-week honeymoon in Portugal and Spain. As part of the trip we wound our way down through Portugal to the town of Sagres, which is on a peninsula that juts out into the Atlantic Ocean, Cape St. Vincent; it's the southwesternmost tip of the country and the whole European continent. It's an incredibly special place, with 250-foot cliffs that plunge down into the sea, and for a long time people believed it was the end of the world. A Portuguese prince, Henry the Navigator, had a school there; you can still visit the ruins. He had the latest nautical charts and celestial maps and navigation equipment. It's where explorers would meet and share knowledge and information before setting sail across the ocean. Being there made me think about Columbus. Magellan. Vasco da Gama. The guys who took great leaps into the unknown to discover new worlds. Sagres was the Johnson Space Center of its day.

During the day we visited the ruins of Henry the Navigator's school, and that night we went to an outdoor concert in Sagres, out on these cliffs that plunged 250 feet into the sea. It was one of those perfect nights, listening to the waves crash, feeling the warm summer breeze, watching the sun set over the ocean as the moon rose up in the sky. I looked out over the cliffs and I thought about the explorers who had sailed from places like this, what they'd accomplished, mapping the known world, charting our place in the universe. How many times had they failed and fallen down only to get back up and try again? How many times had they sailed out on an impossible voyage and made a successful return home? I sat there with Carola looking out over the endless horizon. It was strange, but I felt like everything was going to be okay. The end of my story was not yet written, and I still had the chance to make it extraordinary.

6

HUMAN FACTORS

That fall, while Carola and I were unpacking boxes in our apartment on Massachusetts Avenue, I got a call from a flight surgeon at NASA with a question about my application. I knew right away why he was calling.

"Hey, what's the deal with your eyes?" he asked.

This was the moment I'd been trying to avoid. On the astronaut application there's a box where you're supposed to put your eyesight down if you know it. I'd left the box blank, thinking they'd see it and go, "Oh, maybe he's never had his eyes checked, so he must see fine." But the reason I'd left it blank is because I'd had them checked, many times, and I knew they were bad.

I'd known since seventh grade when I was in the stands at a Mets game trying to write down the lineup on my scorecard, and I couldn't read the scoreboard across the field. I got glasses, but I hated them so I didn't use them that much. I tried wearing them during a baseball game once, and I took a line drive that broke my nose. After that, I went around blind most of the time. By eleventh grade my eyes were bad enough that I had to squint to see the basket on the basketball court. I started wearing contact lenses, and from then on I was fine. Until NASA called.

Since I didn't come from a military background, I didn't know how important good, uncorrected eyesight was for becoming an astronaut or a test pilot. As it turns out, it's very important. It's a deal breaker. If you don't make the cut-off, that's it. Done. Finished. You're out.

The flight surgeon didn't know I'd skipped the question. He thought I'd missed the box. I fished out one of my old prescription slips and told him what it was: minus 3.5 diopters on one eye and minus 4.0 on the another eye. He said, "Yeah, that's no good. That's about 20/350 or 20/400. We don't need 20/20, but we do need at least 20/200. We can't take you."

I said, "Is there anything I can do?"

This was back before LASIK or any of these other fancy procedures they have nowadays. There was a surgical procedure called radial keratotomy that people did back then, but NASA still wouldn't accept you with that; they didn't trust it. "There is this thing you can try," he told me. "It's called orthokeratology. Check it out. Maybe you can give that a shot and resubmit another application if you can get your eyes better. But based on what we have now, I'll have to reject you."

Then, as we were getting off the phone, he said, "Look, if your application got to me, that means you're in the highly qualified section of candidates. You're in the top ten percent, and you should feel good about making it this far. If you can get your eyes fixed, you've got a real chance."

When I hung up I was in shock: the top 10 percent! That was all the motivation I needed to keep going. I wasn't even that bummed about the eye thing. It was another obstacle I had to deal with, but maybe I could fix it. I was ready to take on anything if I was that close to making my dream come true.

I did some research and I learned that orthokeratology is a process where they use contact lenses to reshape your eye. When doctors first started prescribing contacts to people back in the 1940s, the first lenses they invented weren't the nice, soft ones we have today; they were pieces of hard glass or plastic. Doctors noticed that, after wearing these hard lenses for a while, people would wake up and see without any assistance.

The eyeball is a lens. Light comes in, hits the lens, and gets bent to hit the retina on the back of the eye. If it hits the retina at the correct angle, you see 20/20. If it doesn't, you're either nearsighted, farsighted, or astigmatic, and you get glasses or contacts to bend the light at the correct angle so you can see. What these hard lenses would do was reshape your cornea—they'd flatten your eyeball, basically—and you could see. The problem was that, once you took the lenses out, after a couple days your eyes would go back. The tissue pops back to its natural resting place. But supposedly, if you stuck with it, you could get your eyes to see better unaided for a while.

I decided to give it a shot. I hauled out my humongous *Greater Boston Yellow Pages* and found a doctor who specialized in orthokeratology. He prescribed me these hard lenses, and my vision started getting better. It would stay better for a couple of days after I took the lenses out. After a few months my vision fell within the standard for the astronaut program and I was ready to submit another application. Surely NASA would want me now. All I had to do was go back to MIT and do the impossible thing that had nearly killed me: pass my qualifying exam.

. . .

Classes started up again at the beginning of September. My next crack at the qualifier was scheduled for the end of November; Professor Sheridan was on sabbatical at Stanford that fall, but he'd be home for Thanksgiving and that was the only time he could do it. I had three months to turn everything around.

I took statistics that fall, and I made a buddy in that class, Roger Alexander from Trinidad. We bonded quickly because we both fell into the hardworking-regular-guy category and not the eccentric-supergenius category. Roger lived in one of the graduate dorms, which was like an apartment with four guys sharing a kitchen and a common area. We'd study there in the evenings. I'll never forget one night when we had a problem set that was killing us. We couldn't crack it, and then, around two in the morning, Roger's roommate, Greg Chamitoff, wandered out into the common area in his underwear, eating this gigantic orange. Chamitoff was one freakishly smart dude. You could tell there was a lot of power in his brain. MIT's mascot is the beaver, because the beaver is the engineer of nature. Not coincidentally, it also does most of its work at night. Chamitoff fit the description cold. I never saw him much during daylight hours. He walked over to us, looking like he'd just woken up, and said, "What are you guys doing?" We told him this problem had us stumped. He asked if he could see the textbook, and we showed it to him. He looked it over for about a minute or so and said, "Yeah. Do this and this and this, and you got it." We'd been staring at this problem all night, and Greg solved the whole thing in his head in a few seconds, standing there in his underwear, slurping orange juice off his fingers.

Greg wanted to become an astronaut, too (and he eventually would, a couple years after me). At that point he'd already passed his qualifier. We started talking. He told me what he did to pass was that he pre-practiced the oral exam with his friends. They would grill each other like they were each other's thesis committee, because passing the exam wasn't just about knowing the information—it was being able to anticipate the questions and think on your feet without getting rattled. Greg offered to do the same for me: assemble a practice committee of guys who'd passed their qualifiers, who knew what the drill was. They'd put me through the paces and toughen me up.

Talking to Greg that night I realized I'd prepared for my oral exam completely wrong. I hadn't asked anyone for help, so I hadn't known what I was doing. Sitting alone in my study carrel, cramming my head with facts and information without learning how to do the actual thing itself, it was no wonder I'd failed.

We have this idea in America of the self-made man. We love to celebrate individual achievement. We have these icons like Steve Jobs and Henry Ford and Benjamin Franklin, and we talk about how amazing it is that they did these great things and built themselves up out of nothing. I think the self-made man is a myth. I've never believed in it. I can honestly say that I've never achieved anything on my own. Whether it was my parents encouraging me to follow my dreams, or mentors like Jim McDonald who saw something in me, or classmates like Greg Chamitoff who challenged me to do better, I owe everything I've ever accomplished to the people around me—people who pushed me to be the best version of myself.

That's what I responded to when I saw *The Right Stuff*: the people, the camaraderie, the way that John Glenn and the other guys stood up for each other and looked out for each other. It wasn't just that I wanted to go to space—I wanted to be a part of the *team* that went to space, because they seemed like a great team to be a part of. That's why I fell in love with sports, too. Being a part of a club, having that fellowship, it's where I feel at home. I think I liked the friendship and the camaraderie of sports more than the actual playing.

The mistake I made with my PhD was that I forgot to find a team. I thought I was running a marathon by myself, and that's how I'd trained for it. I took Greg up on his offer to run the mock oral exams. Every week my fellow students would grill me. Nick Patrick, a British guy who ended up becoming an astronaut in the class after me, helped a great deal. So did two other guys, Cliff Federspiel and Mohammed Yahiaoui. They were merciless. Every week I'd stand up there alone at the blackboard with my little piece of chalk and they'd tear me to shreds. They made me rethink the weak assumptions I'd put into my work. They made me learn how to think on the fly and express my ideas clearly. They'd pound me and pound me and pound me. Then we'd head over to the Thirsty Ear for a drink.

What amazed me was that Greg and the other guys didn't have to help me. They were carrying full doctoral course loads, too, and they'd already passed their qualifiers, so it's not like I was doing anything to help them in return. But they did it anyway. Because that's what you're supposed to do. That's how a team works. You help the people around you, and everybody's better off for it. The crazy thing is that most of those guys wanted to be astronauts, too, but they never saw it as a competition. We were on the same team, where you want everyone around you to be as successful as possible, because in some way or another their success will become your success. It's good karma—what goes around comes around.

When Thanksgiving week rolled around it was time for me to face the real firing squad again. On Wednesday morning I went back down to Sheridan's office. It was the same setup as before, my advisors seated around the coffee table, me with my little piece of chalk standing in front of the chalkboard. They got settled in and went to work on me.

They hit me with a ton of tough questions in a row. *Bam, bam, bam,* switching back and forth from control systems to spacecraft systems to neuroscience, jumping around to try and trip me up. Then they came after me on my research. The first time I went in my research wasn't solid. Now, thanks to my weekly grilling from Greg and his crew, I'd thrown out my weak ideas and assumptions. My work stood up and I was able to defend it. The whole experience was every bit as brutal as before. I didn't breeze through it by any means, but I wasn't stumbling and stuttering through my answers. I stayed calm and focused and on my game for the whole two hours.

When it was over they asked me to leave the room so they could decide my fate. I stepped out and closed the door behind me. Then, quietly, I turned and put my ear to the door to try and hear what they were saying. I heard one of them say, "Well, he's obviously got a lot of skills, but . . ."

When I heard that "but" I turned and walked away. I didn't want to hear what came after that. I left the building and walked around campus and tried to ignore the questions racing through my mind. Did I pass? Did I fail? Am I staying? Am I going? Am I going to have a good Thanksgiving or a crappy Thanksgiving? Whatever happened in that room was going to alter the course of the rest of my life.

I knew, walking around, that there was still a decent chance that I'd failed. Weird as it sounds, I was ready to fail this time. I'd failed the first go-round because I wasn't prepared and had made a bunch of stupid mistakes. That I couldn't live with. But if I failed this time, at least I'd know I went down swinging and giving it everything I had. If you're going to fail, that's how you want to do it.

After a half hour I went back to Sheridan's office to get the news. They started off with this long list of things I needed to work on. You did well on this, but you need to work on that, all very vague. I was good in control systems, Sheridan said, but I needed help in basic engineering concepts. Then one of them said maybe I should be a TA, that teaching undergrads would help me work on the areas where I needed help. That set off this whole discussion. "Maybe he could TA such-and-such course." "No, I think maybe he should TA this other course." And on and on.

They kept going back and forth with each other in their absentminded-professor way, talking about me like I wasn't even in the room. I *thought* they were saying I was still a student at MIT, but I couldn't actually tell. None of them had actually said the words "You passed." This dragged on for what seemed like forever. It was driving me crazy not knowing. Eventually I jumped in and said, "Um, can I ask a question? I'm going to go home for Thanksgiving tomorrow, and my mother is going to ask me if I passed my qualifying exam. What do I tell her?"

Sheridan stopped and gave me this look. He said, "Oh. No, no. Yeah, you passed. We're just trying to figure out what you need to do next."

That was all I needed to know. The rest of the conversation I nodded and smiled and said yes to everything. I told them I would teach whatever they wanted me to teach, I would take whatever they wanted me to take. I didn't care: I'd *passed*.

The first thing I did was find a telephone. I called my wife. I called my parents. Everyone was thrilled, but I think they were more relieved than thrilled. They said "Thank you, God" and

"Maybe now we can finally have some peace around here." That night Carola and I drove down to New York to see our families for the holidays, and they had a champagne toast waiting for us. When I went to bed that night, I was actually afraid to go to sleep for fear that I'd wake up to realize I'd dreamed the whole thing.

If you work hard and get help from good friends, together you can overcome almost any challenge, no matter how great. More than aerospace systems or neuroscience or anything else I studied, that life lesson was the most valuable thing I learned at MIT. And as I pursued my dream, long after I became an astronaut and even when I was floating by myself 350 miles above the Earth, it was a lesson I would return to again and again and again.

7

DISQUALIFIED

At 10:00 p.m. on March 16, 1993, Carola gave birth to our beautiful baby girl, Gabby. We took her home and it was like this light had come into my life. Everything seemed better. The air smelled better. The trees seemed more beautiful. People sometimes think having a kid gets in the way of pursuing a dream. I think it's the opposite. Having her made me want to pursue my dream even harder because I wanted her to be able to do the same. I didn't want to tell her about how to live life—I wanted to show her.

Gabby was born in Houston. Eight months earlier, after finishing my PhD, I'd taken a huge gamble and moved to Texas for a job with McDonnell Douglas. The chance of my becoming an astronaut was still the longest of long shots; I'd applied to the program a second time in the summer of 1991 and been rejected again. But I believed that being a part of the aerospace community in Houston, where I could be close to the space program and get to know the people involved, was the best chance I had. I reached out to Bob Overmyer at McDonnell Douglas again, and the company made me an offer to head up their independent research and development team for robotics. With big government contracts, part of the budget is often earmarked for research. The whole purpose is to think big, generate new ideas, run experiments, get published in scientific journals. My job would be to think up new ways to use robots in space. On August 19, 1992—my thirtieth birthday, as it turned out—my newly pregnant wife and I packed up our apartment in Boston and started a new life. In Texas.

I don't deal well with transition, and starting a new job in a new state on the day you turn thirty while your wife is expecting is a pretty mind-blowing transition. I was leaving the grad school bubble behind and entering the real world. Real job. Real family. *This is it. There's no going back.* Throughout the move doubts were swirling around in my mind, and I couldn't shut them up. I thought I was going to be an *astronaut*? Who was I kidding? At the time, the whole thing felt like a horrible mistake. Then Gabby was born, and having her brought life into focus and reminded me why I was doing what I was doing.

We bought a house in Clear Lake, the suburb southwest of Houston where the Johnson Space Center is located. After a lifetime in the Northeast it was a rough adjustment, but we'd stumbled into a wonderful community and we slowly got acclimated. Clear Lake is a company town. Nearly everyone is tied to NASA and the aerospace industry in some way or another. Our neighborhood was right off Space Center Boulevard, about five minutes from the entrance to the Johnson Space Center. It was like living in Astronautville. My whole life these guys were my heroes. Now they were my neighbors.

Steve Smith became an astronaut with the class of '92 and was on his way to becoming one of the top spacewalkers in NASA history. Steve is one of those people who's always in a good mood, has a huge smile, is friendly to everyone. He was so generous there were times I thought he wasn't human. He was also tall and impossibly fit. Those barbells at the end of the rack that are covered with dust because nobody at the gym uses them? Steve would go right for them. He was an All-American water polo player at Stanford and captain of the 1980 NCAA Championship team. He's one of those guys who's phenomenal at everything—but you can't hate him for it, because he's also the nicest guy you've ever met. Steve lived right around the corner and had a daughter about Gabby's age. He became a close friend, mentor, and confidant.

I started running into astronauts everywhere I went, even at our church, St. Clare of Assisi Catholic Church, which was so new it didn't have a building yet. While that was being built they held Mass in a storefront in a strip mall next to a hardware store. I called it St. Clare of the Shopping Mall. Kevin Kregel and his family went to St. Clare with us. He was a fighter pilot out of the Air Force Academy who did an exchange with the Navy to attend the Navy Test Pilot School. Better than that: Kevin was from Long Island. The first time we met, he'd heard me speaking and walked up to me with a raised eyebrow. "You ever go to Solomon Grundy's?" My eyes lit up. Solomon Grundy's was a rock club on Long Island in the eighties; I loved the place.

"Yeah," I said. "You from around there?"

"Yup," he said. "I placed you right away with the accent." Kevin was a bit older than me and already had four kids, but he could easily have been one of the guys on my old Police Boys Club baseball team. Knowing that someone who grew up minutes away from where I did had become an astronaut was a huge inspiration.

Working at McDonnell Douglas, I was back in the sea of beige cubicles again, back in a white shirt. It was everything I'd run screaming from at Sperry and IBM, only now I was in a different world. I was right down the road from the Johnson Space Center and Ellington Field. I can still remember my first Saturday living in Clear Lake. The Texas Air National Guard flies F-16s out of Ellington, and one of them came screaming overhead. Most homeowners wouldn't care for that, but I thought it was awesome. I was that much closer to where I wanted to be, and that made everything worthwhile.

As for the job itself, I wasn't sure I'd like it at first, but in the end, it turned out to be the perfect stepping-stone. For my R&D lab, I only needed one idea, like a thesis, something big to sink my teeth into that could get me published in scientific journals. I also wanted to design and build something that NASA needed, something that would make a real contribution to human spaceflight. I convinced the people up at the Johnson Space Center to let me go through some robotics training and work with some real astronauts who could help me understand how the shuttle's robotic tools could be improved. One afternoon in the spring of '93, right around the time Gabby was born, I was standing up on a simulator platform, where some astronauts were training to "fly" the shuttle's robot arm. The arm's official name was the remote manipulator system, or the RMS. Since it was made in Canada, we also called it the Canadarm. It was a giant crane used to move objects around outside the shuttle, like satellites or space station modules. It was also used to position spacewalkers to perform their tasks; they would ride on the front of the arm in a foot restraint. In the microgravity of space, the arm can manipulate something with the size and mass of a Greyhound bus.

The arm was controlled by astronauts inside the shuttle while they looked outside into the payload bay of the orbiter through the aft flight deck windows. They manipulated it via two hand controllers: a left-handed one for translations (XYZ motions), and a right-handed one for rotations (roll, pitch, and yaw motions). Flying the arm required a fair amount of training and skill and was one of the major jobs an astronaut performed on the space shuttle.

As I observed this simulation, I noticed they were using cameras to track the arm's movement but they didn't always have a clear view with the camera, so they were looking at digital readouts to get the arm's XYZ coordinates or its pitch, yaw, and roll. Then they were taking that data and figuring out what they needed to do on the fly. It was an incredibly convoluted and counterintuitive way to manipulate this arm. It was similar to the sensory feedback issue I'd dealt with at MIT. The control system for this robot arm needed better human factors.

I realized that the solution to this, what they needed, was a visual display that rendered the data graphically in real time, like a video game. There was another engineer at McDonnell Douglas at the time, Jack Brazzell, who had figured out a way to help shuttle rendezvous by getting data to display graphically through a software interface on a laptop computer. At that time, if you wanted to introduce something new for the crew, you'd put it on a laptop. When it came to changing or adding software to the shuttle's onboard computers, NASA was very conservative. Those systems were so fine-tuned that you didn't want to mess with them unless you absolutely had to, so they rarely did. Laptops changed everything. They allowed astronauts to make some forward-thinking innovations because now they were free to experiment and try new things. I started working with Jack, piggybacking off some of the work he had done on his laptop rendezvous display, getting his advice on how to get my project on board the shuttle. I found a couple of great programmers, Albert Rodriguez and Mike Meschler, and I brought them on my team to help me create this laptop video game interface to improve control of the robot arm.

Within a few months we had a working demo, and then, much to my surprise, I found my sales experience at IBM coming into play. Because now I had to sell my idea. I had to demonstrate to astronauts and others at NASA the benefits of the new system. If you want NASA to adopt something you've built and incorporate it into their program, you need people—astronauts, especially—to get behind you and tell the decision makers, "Hey, we want this. We need this."

McDonnell Douglas had strong relationships with the astronaut office already in place, of course, and I used those to start knocking on doors. I demonstrated my display system to whoever would listen. I got some polite rejections from a few people who didn't get it, but I finally connected with a Swiss astronaut named Claude Nicollier, the chief of the Astronaut Office Robotics Branch. Claude was a former fighter pilot, tall and thin with an air of European sophistication, but warm and gracious and quick to make a joke. He spoke perfect English with a slight, very elegant accent. The first time I met him I was getting my display hooked up to a simulator. Claude walked in eating a vanilla ice cream cone. He stood there quietly eating the ice cream in his elegant, European way, and watched me try the display. "I like what you have," he said. "We should talk more." He sounded like a Swiss James Bond.

Claude started talking up my idea, and people started to take notice. Jan Davis and Ellen Ochoa, who worked with Claude in the robotics branch, both loved the display and started helping me design and implement it. Jan was from Huntsville, warm and friendly and down-to-earth. She grew up in the shadow of the Marshall Space Flight Center and, like me, had dreamed of going to space for as long as she could remember. She'd actually gone to the launch of *Apollo 11* and made a sign that said, LOOK OUT, MOON. HERE COMES HUNTSVILLE.

Ellen Ochoa's flight crew from STS-96 nicknamed her "E. F. Ellen," after the slogan from the E. F. Hutton commercials: "When E. F. Hutton speaks, people listen." She was actually quite soft-spoken, but she was so sharp and capable that she could command a room because everyone knew that her thoughts were valuable. To no one's surprise, she went on to become head of the Johnson Space Center.

Jan, Ellen, and Claude started pushing to get my robot-arm display flown and tested on a future shuttle mission, and together they made it happen. NASA eventually decided that my display would be flown and tested on STS-69 in June 1995. It was one of the proudest moments of my life to that point: Even if I never made it to space, something I'd created actually would.

Working with the people in the robotics branch the way I did made me thankful once again for my own dumb luck. If I'd had a clue what I was doing as an undergrad, I might have been a more traditional aero/astro guy and specialized in something like jet propulsion, and I'd have been off in a lab somewhere working with a bunch of machines every day. But I hadn't specialized in jet propulsion. I'd specialized in human factors, which meant I was working with the humans—the astronauts. In hindsight, it was one of the best calls I could have made. Because when it comes time to choose a new class of astronauts, for the most part the astronauts do the choosing. No politician in Washington has any say in who gets to fly in space. Astronauts make up the majority of the votes on the selection committee, and through my work I was getting to spend hours and hours getting to know them.

And I liked them. It was the same with the astronauts I became friends with outside of work, guys like Kevin Kregel and Steve

Smith. The more and more astronauts I met, the more I realized that they were my favorite people of all time. They were good people. Smart, dedicated, generous, decent. Everyone was happy to go to work every day. I'd had this fantasy of what astronauts would be like from watching *The Right Stuff,* the sense of camaraderie and shared purpose. And it turned out that was the reality; if anything, reality exceeded my expectations, and every time I drove in to the Johnson Space Center I had a voice in the back of my head telling me: This is it. I want to be a part of this. I want this more than anything.

The astronaut selection process takes almost a year. In the summer of '93, NASA started taking applications for the astronaut class of 1994. I submitted mine. Then there was a problem with funding and the class of '94 got scrapped. They held on to everyone's applications and told us they were going to wait a year and pick people for '95. A year ticked by, and the following summer I updated my résumé and my recommendations, sent them in, and waited by the phone.

I was an old hand at this by now, no longer flying blind. I had people to talk to, and they showed me how the process works. One of them told me, "You do know you can see your file, right? You can file a Freedom of Information Act request and see what they've got on you and what they've said about you in the past."

Smart idea. I requested my file and, sure enough, I saw a mistake I'd made, why my application had hit a wall the second time. When I'd worked at NASA headquarters in the summer of '87, my supervisor was this guy who was a little aloof. We weren't close, but I'd put him down as a reference anyway because he was a big name. I did that with a couple of the recommendations: I picked people I thought were important instead of people who knew me. That was a mistake. This supervisor and I, we hadn't interacted much, and he'd checked off "average" or "don't know" to nearly every question. At one point he'd checked *Don't know, don't know, don't know* all the way down the page. The last question was open-ended: "What else can you tell us about this person?" He'd written in, "Dammit, I don't know."

So that was bad. I wasn't going to make that mistake again. I asked Ellen Ochoa to write me a letter of recommendation, and she did. At McDonnell Douglas I had Bob Overmyer. He knew me well and wrote me a great recommendation. I had my PhD, I'd published several papers, I was building this display for the shuttle. It had taken me ten years. Exactly one decade after walking out of *The Right Stuff* at the Floral Park theater, I'd put together an astronaut application about as strong as I was ever going to get.

On August 4, I got the call. It was a Thursday. I was sitting at my desk at work and the phone rang. A woman's voice said, "Hi, I'm Teresa Gomez from the astronaut selection office. We're wondering if you would be interested in coming in to interview to be an astronaut candidate?"

"*Yes*," I said. I was practically jumping up and down.

She said, "Okay, this is a bit of short notice, but we have someone who canceled for next week, and we're trying to get someone local so we won't have to arrange travel. But if you can't make it next week, you can wait and come in week five or six."

"I'll come next week," I said. "I don't want to wait and take any chance that you'll change your mind."

She said, "Don't you need to see if you can get off work?"

I said, "I'm coming. I'll quit my job, I don't care."

"That may not be a smart thing to do."

"Okay. I'll ask and call you back."

I checked with Bob Overmyer and Mike Kearney, and of course they were supportive. They were rooting for me. Places like McDonnell Douglas actually want you to leave and become an astronaut. It's a feather in their cap. I called Teresa back and told her I was in, and she had me come by the Johnson Space Center to pick up an information packet. I went home and read it, and it was fairly basic: where to be, how to dress, etc. Then I came to the part about the eye exam. It said, "You will be given a series of intensive eye exams. We ask that those of you who wear contact lenses do not wear them for two weeks leading up to the test." When I read that, I knew I was in trouble.

They insist on that because they want your eyes as close to their natural state as possible for the exam; contacts can cause edema, a swelling of the eye, and they want you to be completely free of that. But I was going in on short notice. I didn't have two weeks to let my eyes rest. And I was still using those orthokeratology lenses to flatten my eyeballs. I'd been wearing them for four years now. I knew if I took them out, within a couple of days I wouldn't be able to see anything. I called my eye doctor and he said, "Your lenses won't cause any edema. I wouldn't worry about it." So I thought about it and decided it would be okay for me to leave the contacts in.

At that time, NASA got anywhere from three thousand to five thousand applications per class. Out of that, they go through everyone's qualifications and check references and cull it down to 120 semifinalists. Over the course of six weeks, they bring those people down to Houston in groups of twenty for interviews and medical evaluations. From that pool they make their short list, anywhere from ten to twenty to twenty-five, depending on what they need that year. Then, if you make that cut, they put you through an exhaustive background check. Out of the people who pass the background check, the committee makes its final selections.

To start off, you show up Sunday morning for a week of tests and interviews and evaluations. The other nineteen candidates in your week's group are there. It's like a Hollywood audition where everyone's there for the same part; you're all friendly and professional, but everyone's quietly sizing up the competition. First thing you go through is a lot of written tests: IQ tests, personality tests. There's an ethics test with weird questions like "It's okay to kill someone if . . ." Then over the course of the week you have a series of medical exams. They check you out from top to bottom. You're like a lab rat. They look in your ears, in your throat. There are brain scans. CAT scans. Blood samples. Urine samples. Stool samples. They do an ultrasound of your internal organs, looking for tumors or aneurysms. They stick a camera up your rear end and check in there, which was new for me. On one day they give you a heart monitor to wear for twenty-four hours to track you for any irregularities in your heart. They're looking for anything they can find wrong with you anywhere. By the time it's done you've been poked, prodded, and picked over in ways you didn't know were possible.

For the psych evaluation, you have to sit with two shrinks for a few hours and talk about your mom and dad to make sure you can handle everything mentally and emotionally. They also put you through different stress tests. In one of them, to see how you deal with claustrophobia, you're zipped up inside this dark canvas bag, a personal rescue sphere. They leave you in there and don't tell you how long it's going to be. I fell asleep. That was easy. Then came the part I'd been dreading: Thursday morning, August 11, I was scheduled for my eye exam. I kept my orthokeratology lenses in until the appointment, took them out, said a prayer, and went to meet my fate.

NASA had two optometrists, Bob Gibson and Keith Manuel. Their job was to conduct the eye exams and report the results to the flight surgeons who made the final recommendations on medical fitness to the selection committee. Rainer Effenhauser was the flight surgeon overseeing my group of applicants, but Smith Johnston was another flight surgeon on staff I got to know, and he helped talk me through parts of the process as well. Keith Manuel was running the eye exams that day. He ended up being my neighbor, a great guy, but that day I walked in hating him. I knew the eye exam was the biggest thing between me and a clean bill of health, and I went in thinking of him as my nemesis.

The first thing Keith had me do was the standard eye chart to test my unaided acuity and to see if he could correct me to 20/20. He put me through the paces. "Read line one." "Read line two." He started using different lenses to try and correct me to 20/20, and he got frustrated. He couldn't do it. He said, "I don't know what the problem is. You're not seeing 20/20 no matter what I do here."

Then, to test my unaided acuity, he put me on the Landolt C machine. I'd never heard of it before, but it's way more sophisticated than the eye chart. It's a machine that flashes the letter C in front of you over and over again, randomly, in rapid succession, projected at different depths and with the open part of the C facing in different directions. It's fast. You can't sit there taking your time. You have a joystick and you have to move it up, down, right, or left in response to which way the C is facing.

They say the Landolt C is the most accurate vision test ever created. Needless to say, I did not pass. It was a disaster. I can't remember how many I missed; it might have been all of them. Finally

Keith said, "Okay, the last thing I have to do is map your eyeball." Map my eyeball? I didn't even know what that was. With all the crazy things I'd had done to my eyes, no one had ever brought it up before. Basically it's exactly what it sounds like. They use this machine to take a 3-D topographical map of your eyeball; it's a way to determine the shape and health of your corneas. Keith hooked me up to this contraption and futzed with it for a few minutes. He said, "You've got a flat eyeball."

"That's from the orthokeratology," I said. "One of your flight surgeons recommended that to me as a thing I could do."

"Yes," he explained, "but orthokeratology lenses are a special case. You were supposed to stop using them six months in advance to let your eyeball revert to its normal shape."

I hadn't known that. I wouldn't get the official results until the next morning, but I knew they'd be bad. Even worse, I still had the most important part of the week ahead of me: the interview. I had to sit down at a conference table with the entire astronaut selection committee. Some of the biggest names at NASA were on this board. John Young was the head of the selection committee. An Apollo moonwalker, commander of the first space shuttle flight, and one of my all-time personal heroes, Young was a character, a true original. He had a thick Southern drawl, and he used it to say exactly what was on his mind. He'd gotten in a bit of trouble when he smuggled a corned beef sandwich onto *Gemini 3* in 1965 (still the world's first and only corned beef sandwich ever flown in space). Young was at the head of the table. Seated around him were Hoot Gibson, the head of the astronaut office; Ellen Baker, an astronaut from New York; Steve Hawley; Brian Duffy; Tom Akers; Duane Ross; and a few other people. I had to sit down across from them and put my best foot forward and sell myself, knowing full well that I'd probably already bombed out.

In the end, strangely, I think it worked out in my favor. I walked into that conference room with nothing left to lose. I couldn't blow it in the interview because I'd already blown it with the eye test. I decided to go in and do my best and let the chips fall. There was a professor I knew at MIT, smartest guy in the world, the top of his field, and I remember the first thing he had listed on his résumé was "Astronaut Candidate Finalist." He never became an astronaut, but he considered the fact that he'd made it that far to be the most important achievement in his career. I'd made it that far, too. Whatever else happened, I knew that these people considered me good enough to sit at that table with them.

I'd talked to my astronaut friends about how to handle the interview, and every single one of them told me, "Just be yourself." You don't have anything left to prove at that point. They want to get to know you and see who you are. I'd run into Kevin Kregel outside church the Sunday before and told him about the interview. He said, "Don't try and BS anybody. Don't make anything up. If they ask you something and you don't know, say, 'I have no idea what you're talking about.' Because that's the right answer."

The interview went well. We had a friendly chat, and it was great. They asked me about what it was like growing up on Long Island, my playing the trumpet in the school band, my dad being a fire inspector, random things. They asked a few questions about my work and my research, but mostly it was get-to-know-you type questions. I told a few funny stories, got a few laughs. We got so wrapped up in talking that the hour flew by. Finally, Steve Hawley said, "Hey, we're out of time. Is there anything you want to add?"

I said, "No, just that I appreciate the opportunity. This has been the highlight of my life coming in here, and whatever happens happens." We stood up and everyone shook my hand. Everybody was happy and smiling. I felt good about it. I felt like I belonged in that room. Those were my people. This was the team I was supposed to be on. But I knew that when I woke up the next morning it was going to be the worst day of my life.

On Friday, August 12, 1994—before I even got to the news waiting for me at NASA—Major League Baseball went on strike. They'd played the last game of the season the night before, and then that morning the players of every single team walked out in protest over the salary cap. The rest of the season wound up canceled, and there would be no World Series for the first time since 1904. The strike was bitter and it was ugly and the whole future of the sport looked grim, much like my chances of being an astronaut. It was the mother of all bad omens.

On the drive over to the space center I was actually hoping for something else to be wrong other than my eyes. I was praying for an aneurysm or a tumor, something so far out of my control that I could throw my hands up and say, "Well, that's life. Nothing I can do." No such luck. I was clean as a whistle. My organs were good. My rear end checked out. My hearing was pitch-perfect. My psych test came back with good results: 100 percent sane. They said I was off the charts for happiness. I was basically a happy person who got along well with almost anyone, good traits for an astronaut. I had met and surpassed all the medical criteria for the job—except one.

I sat down with Rainer Effenhauser and he gave me the news about my eyes. "Your unaided acuity is beyond our limit," he said, "so we have to DQ you on that. We couldn't correct you to 20/20, either, so we have to DQ you on that. And you've got a flat eyeball in your head. We've got to DQ you on that, too. I'm sorry, but we can't take you. With these results, there isn't a chance you can be considered. You're medically disqualified."

The words hung there in the air: medically disqualified. Not "underqualified" or "in need of more experience" but physically and genetically unfit for service. I was crushed. It'd been ten years. Ten years of my life I'd been working toward this goal. I didn't know whether to feel angry or sad or frustrated or what. My whole body was numb.

After I got the news, I called up Duane Ross, the head of the Astronaut Selection Office, and asked if I could come by and talk to him. I wanted to know if there was anything—*anything*—that I could do. Duane had been head of the selection office since the shuttle program began. He was the warmest, most gracious guy, the kind of guy you wanted in your corner because you knew he'd do whatever he could for you. He told me to come by and we sat down and he couldn't have been nicer. He said, "Mike, I want you to know we were all disappointed when those results came back. I can't tell you we would have picked you, but I can tell you that you were one of the people we were talking about. Maybe you wouldn't have gotten it this time, but you might have gotten it in a future selection."

To hear him say that broke my heart: They wanted me. I was so close. It was right there in front of me. I called some of the astronauts on the selection committee and asked them if they minded giving me their feedback, too. They all took the time to speak with me and not one of them said, "Hey, this isn't worth your while. Good luck." If they had, I think I might have given up. But they didn't. Every single one of them took me aside and told me, "You know, if you can do anything about your eyes, you should give it a try."

At that point I decided if I was going to be told no, I wanted to be told no. I didn't want to be told, "We wish we could have." After everything I'd invested, for me to walk away the door had to be closed and closed forever. As long as it was open, even just a crack, I knew that I couldn't bring myself to stop trying. I'd made it too far and come too close to give up, and I had nothing left to lose. There was only one thing I had to do to get back in the mix: I had to learn how to see.

8

YES OR NO

Monday morning I was back at work at McDonnell Douglas and I ran into Bob Overmyer in the hall. He said, "What happened?"

"I got medically DQ'd."

"Your eyesight?" he said. "Yeah, that happens with everybody. I fought those tests for years." As a former pilot, Bob knew all about the eye test. For as long as there have been planes in the sky, pilots and astronauts have been running scared from the eye exam—because you can be the best, most qualified pilot in the world and then get benched for this thing that's 100 percent beyond your control.

Bob was actually encouraging. "This is only your first time," he said. "Don't give up. You'll get another shot and you may get back into this thing." He told me pilots do all kinds of wacky things to try to beat the test. "You know what I used to do?" Bob said. "I'd dehydrate myself. I'd schedule the exam for Monday morning, and over the weekend I wouldn't drink anything. I'd run like crazy, get all the water out of my system. That way you dry out the eyeball and make it stiffer and it bends the light better."

"Okay," I said. "Makes sense. I'll give it a shot."

That same afternoon I was at the Johnson Space Center and I saw Kevin Kregel in the hallway. I told him what had happened. He said, "Those damn eye tests. They'll kill you every time. But you know what you gotta do, right?"

"What's that?"

"Drink lots of water. Drink as much water as you can, for days. The morning you go in, don't even take a piss. It'll make your eyeball more viscous and it'll bend the light better."

It actually made me feel better, knowing that I wasn't the only guy who'd been through this and nobody else had a clue what to do, either. Bob and Kevin had both faced the same obstacle and they'd overcome it and both had become astronauts. That gave me hope.

The best advice I got came from my neighbor, Steve Smith. He told me, "You have to look at this like any other engineering problem. You have to collect all the information and data you can, figure it out." He was right. I hadn't been dealing with the problem in the right way. I hadn't been to an eye doctor in two years. I'd lulled myself into believing this orthokeratology thing would be an easy fix, but that had been a way to avoid facing my fear head-on. I would have known more about NASA's stance on orthokeratology if I'd been up-front and asked, but I'd been too scared to bring up the subject of my eyes. I thought I could tiptoe around the problem when what I needed to do was tackle it: admit that I needed help and get help. I went back to JSC and went to see Smith Johnston. "What do I have to do?"

Smith started talking, and I couldn't tell for sure, but I got the impression that he'd spoken with Duane Ross and Duane had told him: "See if you can make this work with Mike. Let's not throw him away over a bad eye exam." The first thing Smith told me was to take the damn lenses out of my head. Not only were they not helping, but because I'd worn the same lenses for so long without getting them checked, they'd gotten old and scratched up and had damaged my eyeball, which was why Keith couldn't correct me to 20/20. So no contact lenses for six months—only glasses, to give my eyeballs a chance to heal.

I also started looking into vision training. Overfocusing of the eye muscles is one of the causes of nearsightedness. Vision training is a program of exercises that teaches you how to relax your eye muscles in order to improve your unaided acuity. It's not a miracle cure, but it can give you incremental improvements, which was what I needed. It just takes time. Smith told me to do that and keep getting checked by my eye doctor and send him the results. If I showed that I could pass, they'd consider admitting me in the next class after this one.

I said, "Okay, if that's what I have to do, I'll do it."

Over the course of the next year, about a million things happened at once. Shortly after my flameout on the eye exam, we found out Carola was pregnant again. With two kids to support and the astronaut dream looking shaky, I had to think seriously about what I would do if it didn't pan out.

I liked the job at McDonnell Douglas, and it had been a great way to work closely with the astronaut office; but if I wasn't going to be an astronaut, I wasn't sure it was what I wanted to spend the rest of my life doing. I had started teaching some classes on the side at Rice University, which has a great engineering school and a long relationship with the space program going back to the start: Rice's stadium is where President Kennedy gave his big speech kicking off the Apollo program in 1962. Teaching brought in a little extra money, and in the back of my mind I always thought if the astronaut thing didn't work out, academia might be my best fallback option. With the astronaut dream up in the air, I started sending out letters and résumés to different schools for full-time professorships. I got interviews with the University of Maine, the City College of New York, a few other places. Columbia, my own alma mater, sent me a nice letter telling me no thanks.

Then I got a call from Georgia Tech. Bill Rouse, a former student of Tom Sheridan's at MIT from before my time, was in the industrial engineering school down there and he'd started a lab, the Center for Human-Machine Systems Research. They were doing work with human factors and control systems and were looking to do more space-related work as well. I flew down and interviewed with them and they offered me a job.

I agonized over the decision for months. Was it a good idea to leave the Johnson Space Center, the focal point for all human space flight? On the other hand, this was a full-time tenure-track position with one of the best engineering schools in the country, and I wasn't being offered anything like that in Houston.

Finally, in December, I came to a decision: I would take the job. Georgia Tech wanted me to start right away, in January. I didn't want Carola to have to move while she was pregnant; she had her doctors and everything in Houston, and we wanted the baby to be born there. Also, my robot-arm display was set to be flown in space on STS-69 in June, and I wanted to be in Houston for that. So I asked if they'd wait until the fall semester. They agreed and let me push my start date to August.

For the next seven months I only had one job: fix my eyes. I found an optometrist who specialized in vision training, a woman named Desiree Hopping. First she gave me a new pair of glasses with undercorrected lenses; they would take the strain off my focusing system and help my eyes to relax. Then she gave me some exercises. There was one where I had to stare at a bunch of marbles spaced out on a string at different intervals, shifting my focus to each one. I had to stare at different eye charts at different distances, the idea being that I would train my eye to relax and focus on an imaginary point beyond where the chart is, causing the letters on the chart to appear sharper. These exercises required deep, deep concentration. I had to do this dead stare for minutes at a stretch, no blinking. I looked like a serial killer giving you the evil eye. Some nights I'd go to bed and my eyes would be bloodshot from the strain of forcing them to relax, which sounds odd but it's true.

I'd go to the office every day and work on my robot-arm display. Then I'd come home. We'd eat dinner, put Gabby to bed. Then I'd sit up at the kitchen table doing these vision exercises. My mother-in-law, who'd come down to help us while Carola was pregnant, would sit there with me, holding up these charts over and over again while I stared her down like a crazy person.

But it was working. I kept going back to Dr. Hopping every two weeks to get my eyes checked, and they were getting better, bit by bit. Then NASA threw me a curveball. After pushing the class of '94 back to '95, instead of waiting the normal two years to do the next selection they were going to move ahead and do two classes back-to-back. They'd be taking applications that summer for a class to be picked in the spring of 1996. I'd thought I was going to have a whole year to get settled in Atlanta and slowly work on my eyes. Now the whole interview and selection process would happen right when I was moving. I resubmitted my application and prayed I'd be ready in time.

We put the house on the market, and 1995 turned out to be one of those years when nobody was buying houses. Our place sat on the market for months. No takers. We lowered the price. Nothing. Lowered it again. Still nothing. My mother told me to plant a statue of St. Joseph in the ground; he's the patron saint of getting your house sold, apparently. So I did. That didn't work, either. My display experiment got pushed back to a shuttle flight in September, which meant I was going to miss the last couple months of working on that. The Mets were 23–41. There was a lot going on.

In the middle of all this, on July 5, our son, Daniel, was born. Having a girl was great, and getting a boy rounded out the team. At that point our lives were up in the air and I was racked with doubt about the choices I was making. Daniel's arrival was just what I needed. It was a perfect blessing. Having those kids opened up a new dimension of love for me that I couldn't have dreamed existed. Whether I became an astronaut or not, nothing was more important than that.

Finally, our real estate agent came by and told us she'd found someone to rent our house short-term. Fine, we said, we'll do that. In the back of my mind I was thinking that if I did get picked we'd be able to come right back, which was nice to contemplate. The movers came and started packing us up, and I flew to Atlanta for a long weekend to race around and try to find us a place to live. Our whole lives were up in the air.

Apollo 13 came out that summer, and since I was by myself in Atlanta, I went to see it. That was a mistake. It was the best space movie since *The Right Stuff*: the astronauts and their families having parties in Houston, the whole NASA team pulling together to save these guys and bring them home . . . it was everything I was leaving behind, and I had no idea if I'd ever make it back. I sat there in the theater, loving the movie and being completely depressed by it. I flew back to Houston, collected Carola and the kids, and the first week of August we loaded up and headed east on I-10 with our whole lives packed into this moving truck that followed behind us. The house I'd found for us to rent turned out to be horrible. It looked okay to walk through, but the foundation was cracked and leaked whenever it rained. There were bugs. Nothing felt right.

Then, around the first of September, we had barely moved in when I got a call from Teresa Gomez in the Astronaut Selection Office asking me to fly back to Houston. They were looking at applications again, she said. Mine was in the good pile, and the flight surgeon said my eyes had improved enough that they were willing to let me come back and try again. There was one catch: I had to fly back in a month at my own expense and do the eye exam. If I passed, I'd be back in the running. If I failed, I was out of luck.

It turned out that my luck that fall was pretty good. I called my optometrist to let her know what was happening and she had some news for me. At the time, there were only two Landolt C machines in use in the entire United States. One was at the Johnson Space Center in Houston. The other one, she had learned, was at the Emory Eye Center in Atlanta. Of all the cities in North America I could have moved to, I'd picked the one with the machine that I needed. She suggested that maybe I call them up to see if I could go down there and use it. So I did. There were two very nice women who ran the Landolt C unit at Emory. I went down and talked to them and asked if they'd let me use their machine to practice. They said yes, and for the next few weeks I went there every chance I got.

The first week of October I flew back to Houston to sit with Keith Manuel and take the exam. He mapped my eyeball and it was healthy. He corrected me to 20/20, and that went fine. Then he had to test my unaided acuity on the Landolt C. With an eye exam they want you to relax. They want to test your eyes at their natural resting state. My problem was that I had to work hard to relax. I had to strain my face to do this evil-eye thing that forced my eyes to relax and focus properly. I was like a kid with a learning disability. I could pass the test; I just needed to work harder to do it.

And I did it. I passed. *I passed!*

Duane called me and told me I was back in the pool of eligible candidates. I could come back and interview the last week of October. I flew back to Atlanta, taught for two weeks, then flew right back to Houston. The whole month I was going all out. I ran every day. I didn't eat an ounce of fat. I watched my blood pressure. I kept my cholesterol down. I wasn't taking any chances. Sunday morning I went in for an intro briefing and took the written psych test. Starting Monday I came back and did the ultrasounds, the camera up the rear end, everything I'd done the last time. On my way out I stopped off to talk to Rainer Effenhauser about something. As I was leaving he said, "We'll see you tomorrow for the eye exam."

I wasn't sure I heard him right. I said, "The eye exam? I already passed the eye exam."

He said, "Yeah, but that was three weeks ago. Something could have changed. It has to be done at the time of selection."

"But I *just* did it."

"No, no, no. That wasn't official. I'm sorry, Mike. This is what we have to do."

I couldn't believe it. It was like a punch to the gut. But it was what it was. You do what you have to do. The following afternoon I went back to the optometrist's office to take the test again, only now it was a different setup. Bob Gibson was working off to the side, and two optometry students from the University of Houston were there as well. Bob and his colleague Keith Manuel were adjunct professors at the school, and students would often work in their office prepping patients and performing tests to get workplace experience.

A young woman administered my eye exam. She had to be about twenty-four years old. My whole life had been leading up to that moment. Everything I'd done, day in, day out, for over a decade, was going to be decided in the next half hour. She mapped my eyeball and put me on the Landolt C to test my unaided acuity. I sat down at the machine and started the evil eye to try to relax and focus. She said, "Sir, you need to relax your eyes." To her it looked like I was straining when in fact I was doing the opposite.

I said "I am relaxing" and kept right on with the evil eye. Because that was how I'd trained myself to do it.

She said it again. "Sir, relax your eyes. We won't get accurate data if you don't. Sir? I need you to relax. Sir?"

She would not stop. She kept raising her voice. Inside my brain I was losing it. I wanted to punch something. I wanted to burst into tears. I wanted to yell, *Lady! You have no idea what's going on here! You have no idea how much time and heartbreak I've been through. This is my whole life at stake. This is my little-kid, playing-in-my-backyard-since-I-was-seven-years-old dream on the line—and you need to shut up!*

Of course I didn't say any of that. I nodded and said "Okay" and did my best to ignore her, and finally she started losing it. "Sir! Sir! Stop the test! Stop it! You cannot do this!" Finally, Bob Gibson heard us and came over and asked what was going on. "He's not relaxing his eyes," she complained. "He's not complying with the test protocol."

Bob stopped the test and took me into his office. He said, "Mike, what have you got the rest of the week? Why don't you come by my office first thing Thursday morning and I'll administer your test myself." He could tell I was frazzled, and he wanted to give me the chance to take the test when I was fresh and my eyes were relaxed. I scheduled a new appointment for Thursday and walked out, feeling pretty upset about the whole thing.

Wednesday I had my second selection committee interview. John Young was back at the head of the table again, but there were several new faces as well. I did okay and I could tell they liked me, but there were still a million reasons why I wouldn't get it: I was up against a completely different group of candidates than last time, they might have hired a better robotics guy the year before, and so on. Also, I could still fail the eye exam and none of it would matter.

Thursday morning came. I showed up at Bob Gibson's office. Walking in, I felt surprisingly calm, at peace. I'd done everything in my power to make myself eligible for the job, and at that point there was nothing else left to do. Whether I passed the eye exam or not, I'd always be able to say that I gave it my best.

That said, I *really* wanted to pass.

Bob opened the door, asked me to come in, and said, "You know what, Mike? If you relax and think positive, I'm sure you'll do just fine." He sat me in the chair at the Landolt C machine and started to administer the test. I took a deep breath and went for it. When I finished Bob showed me the results. He said, "Congratulations, Mike. You did it."

I sat there stunned. I couldn't believe it. I looked up at Bob with tears in my myopic-but-now-qualified eyes and said, "I passed? I can call my wife and tell her I passed?"

He nodded. I think he thought I was going to kiss him.

It had seemed so impossible, so crazy, that I would pull this off, but it worked. It *worked*. It was a miracle. I felt a relief even greater than the relief I felt after passing my qualifying exam. Because passing a qualifying exam falls in the realm of what's possible. Getting your eyes to see better than they normally see is close to impossible. It was proof that no obstacle in life is too great to overcome.

The next day I went by Rainer Effenhauser's office for the results of my other tests. "Everything came back okay," he said. "Now leave. Get out of here before anybody has a chance to find anything wrong with you."

I went home to Atlanta and I waited. At that point there was nothing left to do but think positive thoughts, so I concentrated on getting settled in at Georgia Tech. Being a professor gave me the flexibility to spend more time with the kids. Gabby was starting preschool and I picked her up and dropped her off most days. I took Daniel for walks in his stroller. I made breakfast in the morning and read them stories at night. I missed Houston, but in some ways it was nice not to be in the thick of it while I was waiting. I could keep my expectations in check and not obsess over everything. It gave me a once-in-a-lifetime chance to bond with my kids, an experience I would not trade for anything.

By January I knew I'd made the second-to-last cut. The U.S. Office of Personnel Management started my background check. They call everybody: your family, your coworkers, your kindergarten teacher. No stone is left unturned.

During my interview week I'd gotten to know Mark Kelly. Mark was a Navy pilot, and while we were waiting he and the other Navy applicants had put together this e-mail list to share information about the selection process. Mark was kind enough to put me on it. It was gossip, rumors, speculation. We were all trying to read the tea leaves, desperate for every scrap of information we could get.

April 19 rolled around. It was a Friday. An e-mail popped up from a naval test-flight engineer. She had called Houston to check on something and she'd been told that the calls, good or bad, were going out Monday morning. The second I read that e-mail, I shut down my computer, left my office, and went for a walk. I couldn't sit still. I must have walked the whole afternoon, my mind racing.

It was all I could do to get through that weekend in one piece. I puttered around the house, tried to keep myself occupied, but mostly I obsessed over this call. I was thirty-three years old. For most of that time, the answer to the biggest question in my life had always been "Maybe." On Monday morning it was either going to be "Yes" or it was going to be "No," and either way my whole life would never be the same. Sunday night I went to put Daniel to bed. He'd just turned nine months old, and I can remember looking down at him and saying, "Tomorrow we're going to find out if your daddy is an astronaut."

I took Monday morning off from work. If it was bad news, I didn't want to be crying at my desk at the office. I wanted to be home and ready for the call when it came. So naturally I was on the toilet when the phone rang. Carola came to the door and said, "Mike, it's a guy from NASA."

I ran out, and grabbed the phone. "Hello?"

"Mike? This is Dave Leestma from the Johnson Space Center. How are you doing today?"

I said, "Dave, I don't know. You tell me."

He laughed. "Well, I think you're going to be pretty good because we want you to become an astronaut, and we hope you're still interested in coming."

I said, "Yes! And in case you didn't hear me: *Yes! Yes! Yes! Yes!*"

I was screaming into the phone. Carola started screaming, too. Then Daniel started crying. I think Gabby was confused. When I hung up I still didn't think it was real. I had this panic that maybe they'd called the wrong Mike Massimino. I picked up the phone and called them right back. Duane Ross answered.

"Yeah?"

"This is Mike Massimino again. I just wanted to double-check that you guys made the right call."

"Yeah. Don't worry. We did."

part

3

The Real Right Stuff

9

THERE'S MACH 1

Life changes fast when you become an astronaut. On the day NASA called and offered me the job, I was a university professor who spent his days in front of a chalkboard lecturing a roomful of nineteen-year-old engineering students. Six months later I was breaking the sound barrier in the backseat of a twin-engine supersonic jet.

After getting the call and wrapping up my final semester of classes, at the end of July we packed up and drove home to our house in Texas, waiting for us right where we'd left it. When we pulled into the old neighborhood and up our street, Steve Smith, our astronaut neighbor, had decorated our yard with American flags and streamers and a bunch of signs, which was great. One week later, on August 12, 1996, I reported to the Johnson Space Center for work as an astronaut candidate, or ASCAN. I drove up to the north entrance and flashed my badge, and the guard waved me through. It was the best feeling in the world.

The first week or so was mostly orientation, getting an office, filling out paperwork. NASA put on a couple of social events and mixers for everybody to get to know everybody, which we needed: We were the largest astronaut class in the history of the space program, forty-four of us, thirty-five Americans and nine internationals. Every astronaut class gets a nickname. The original Mercury guys were "the Original Seven," and the second group of nine were called, rather imaginatively, "the New Nine." Once the shuttle era came, the names got more creative: the Maggots, the Hairballs, the Flying Escargot. The astronaut office takes up the entire sixth floor of Building 4 at the Johnson Space Center. The class before us had fifteen people, and space was already tight. Then they had to cram all of us in there. They called us "the Sardines."

Once we arrived, NASA didn't waste any time getting us in the air. Shuttle astronauts fall into two groups, pilots and mission specialists. I was a mission specialist. All mission specialists are trained to fly as backseaters, copilots. It's spaceflight-readiness training. The different shuttle simulators are great, but they're not real. Flying a high-performance jet is as real as it gets. You're controlling a real airplane, working with a real pilot, experiencing real nausea and real turbulence and real gut-dropping, nerve-racking, panic-inducing situations. It trains your mind and your body to feel, react to, and deal with how physically and mentally demanding spaceflight is going to be.

On the first day they measured us for our flight suits. Military pilots wear green flight suits. Astronauts wear blue. The suit comes with the American flag on the left shoulder and the NASA logo on the right breast. Most important, it's got your astronaut's wings. You can get white, silver, or gold. Most civilians get white. I ordered gold. They took a mold of my head for my helmet, traced an outline of my feet for my custom boots. Black or brown? Lace-ups or buckles? Everything is custom fit. They also give you custom leather/Nomex gloves and a watch, a Casio Illuminator on a black rubber wristband with a small NASA insignia, informally called the meatball, on the dial. Then you top it off with the cool sunglasses, Randolph aviators, standard military issue, with straight-back frames with no hook so you can slide them on and off while wearing your helmet—the same ones worn by the pilots in *The Right Stuff*. You can get your call sign printed on your helmet if you want, like you see in *Top Gun*, with guys like "Maverick" and "Goose" and "Ice Man." When I was a kid in school, a lot of kids didn't even know my real name was Mike. Everyone called me "Mass." Then, at grad school and McDonnell Douglas, nobody called me that anymore. The name just went away. But once I became an astronaut people started calling me Mass again. It fit. Soon it was the only thing people called me, and that became my call sign, MASS, printed on my helmet.

To get us ready to fly, NASA shipped us out to the Naval Air Station in Pensacola, Florida, for water and land survival training. Then we were off to Vance Air Force Base for parachute training in Enid, Oklahoma, before heading back to Houston for three weeks of ground school, where we learned the aircraft systems, navigation, FAA regulations, how to deal with inclement weather, flight plans—everything we needed to know to assist the frontseater in flying the jet.

Flight operations for the Johnson Space Center were done out of Ellington Field, which is ten miles up the road going toward Houston. NASA has its own facilities there, a two-story office building attached to hangars for our planes: the WB-57 high-altitude research airplane, the shuttle training aircraft, the KC-135A zero-gravity airplane—the famous "Vomit Comet"—and our fleet of T-38s. Astronauts do our spaceflight-readiness training in the T-38, a two-seat, twin-engine supersonic jet. It can go faster than the speed of sound and cruise at altitude around 700 miles per hour.

Just imagine a Ferrari as a fighter jet. They're small and sleek, with razor-thin wings and a sharp needle nose, painted white with a blue racing stripe and NASA's logo on the tail. It's one of the coolest flying machines ever built.

There's a great story about Bill "Spaceman" Lee, a leftie pitcher who started for the Red Sox back in the seventies. His first day at Fenway Park, he pulled up in his truck and some gruff clubhouse guy tossed a jersey at him and said, "Here's your jersey." Bill caught it and was like "That's it? I'm officially a player for the Boston Red Sox. Shouldn't there be some kind of ceremony or something?"

That's what happened to me the first day I showed up at Ellington. The guys who work for NASA out there are mostly former enlisted guys with lots of tattoos. They pack your chute and check your oxygen and generally make sure you aren't going to die from something going wrong with the plane or the equipment. One of these gruff, ex-military types was working the equipment room that day, a guy called Sarge. He had an enormous mustache, he wore a NASA ball cap, and his shirt was soaked through with sweat. He was chomping on an unlit cigar that looked like it had been in his mouth since Vietnam. I walked in and told him my name. He went through a pile, pulled out my flight suit, and chucked it at me like a used towel. Then he started going through his checklist, piling my arms up with the rest of my gear. I said, "Wait a minute. Let's back up a second. This is my NASA *flight suit*. Shouldn't there be some kind of ceremony or something? Maybe a handshake?"

He looked at me, walked over, reached out, and shook my hand. "There you go," he said, and went back to his checklist.

After picking up my flight suit from Sarge, I went into the astronaut locker room to try it on. All around me I could read the names of my heroes on the front of the lockers. There was a locker for John Young. There were lockers for Jerry Ross and Story Musgrave, two of the greatest spacewalkers in the history of the shuttle program. I'll never forget seeing my locker in there—*my locker*, right next to theirs, my name tag affixed to the front with Velcro. MIKE MASSIMINO, JSC, HOUSTON, with the NASA astronaut wings engraved on it.

I put on my flight suit and slid on the boots. Then the gloves and the watch, finally, in front of a mirror, the super-cool aviators. Putting it all on for the first time felt like putting on a superhero costume. I packed everything up and took it home with me and tried it on again and showed it off for Gabby and Daniel. I must have walked around the house like that for a couple of hours. Fortunately my kids were too young to think I was crazy.

I *was* crazy, though. A little bit. In a good way. I think you have to be to have the drive it takes to get this job in the first place. The more I got to know my fellow astronauts, I found that they were all characters, every last one of them utterly and amazingly unique. We had guys like Don Pettit, who got the nickname "GQ" during a field trip to NASA headquarters in Washington, DC. A bunch of us decided to go for a run before dinner, but Don didn't have any running shoes. So he showed up wearing shorts and black dress shoes with the calf-length socks to match—and he was exactly the type of eccentric genius who could pull that off. Don was a PhD in chemical engineering, a bit of a mad scientist. While serving on the International Space Station, he started growing his own vegetables and even invented a coffee cup that works in zero gravity.

Then you had a guy like Charlie Camarda. When I got back to Houston, one of my neighbors said, "Another new astronaut moved in around the corner. I heard he's from New York, too." I went over and knocked on the door and my life was forever changed when Charlie walked out. Charlie was from Ozone Park, Queens, the son of a butcher, a guy from the neighborhood like me. Charlie had a thick Italian mustache and a swoop of jet-black hair. He answered the door in flip-flops, a pair of shorts, a white tank top, and a gold chain. Charlie was like *Saturday Night Fever Goes to Space*, probably the only astronaut in the NASA gym locker room who wore cologne. But he was a *brilliant* engineer. Absolutely brilliant. Holds seven patents last I checked. Proof that you can take the boy out of New York but you can't take New York out of the boy. We hit it off immediately. Neither of us could swim that well and we flailed our way through water survival training together. To this day he's one of the funniest guys I've ever met.

I loved my fellow civilian egghead PhDs, but thanks to my love of *The Right Stuff*, I also gravitated immediately toward the pilots in my class, guys like Charlie Hobaugh, a Marine pilot nicknamed "Scorch" who served in Desert Storm and flew Harrier jets, the kind that can take off vertically and hover like a flying saucer. One day I was with him and we heard the deafening roar of a fighter jet overhead. Scorch pointed to the sky and looked at me. "Do you know what that is?" he said. "That's the sound of freedom." Yeah, he's that guy. Scorch was also *huge*, completely jacked, the guy at the gym who can do endless, effortless chin-ups while you're struggling just to do two. Scorch was also the nicest person on the planet. You had to be careful what you asked him for, because whatever you asked for he'd give it to you; I was convinced if I asked for his right arm he'd lop it off and hand it over.

Scott Altman, whose call sign was "Scooter," was another pilot who immediately became a fast friend. We also called him the WLA, the "World's Largest Astronaut." Too tall to fly for the Air Force, he became a Navy pilot instead. Scooter had been selected in the class before me but lived four houses down the street, and we were already close. I spent hours of training time in the backseat of his T-38. Scooter was cut from the *Right Stuff* test pilot mold, too.

He flew missions as a strike leader in Iraq, had been awarded just about every naval aviation honor or medal you can name, and he drove a badass blue 1969 Camaro convertible. Scooter was actually Tom Cruise's flying double in *Top Gun*. That scene where Maverick flips his plane upside down and flips the bird at the Russian MiG? That's Scooter. One of my favorite things to do was get in the backseat of his T-38 and pretend to be Goose and act out scenes from the movie.

My first T-38 flight was on October 30, 1996. It was a beautiful, clear autumn day. Carola and Gabby and Daniel came with me to the airfield. Most of the flight instructors out at Ellington were older, flying well into their sixties. They had flown with the Apollo guys, trained Neil Armstrong how to land on the moon. Bob Mullen, the crew chief, had strapped the original Mercury Seven into their training jets. Bob Naughton, the head of flight operations, was shot down over Vietnam, captured, and held in the Hanoi Hilton POW camp for six years. These guys were the real deal. They weren't messing around.

I was going up that day with Frank Marlow, one of the flight instructors. The backseater's main job is to handle the radio and the navigation. You can't land or take off, but you get to do nearly everything else: fly the route, do approaches, acrobatics. The pilot takes you up and shows you how. He flies and then you fly. He demonstrates and you execute and you get to know what you're doing.

That day Frank was taking me out to the practice area over the Gulf of Mexico south of Houston. The way it works is, you go up and then radio air traffic control to let them know you're activating the area. Once you do that, no commercial traffic is allowed in, and it's yours to do whatever you want. It's like reserving a tennis court in the sky. NASA's practice area is called Warning

Area 146-Charlie—about a thousand square miles that goes from 10,000 feet up to 26,000 feet. The Texas Air National Guard, with its F-16s, uses Warning Area 147-Delta for practice; it goes from the water up to 50,000 feet and covers an even larger footprint. There you really have room to have fun.

Of course, the one thing they drill into you is that, yeah, it's a lot of fun, but you're not there to have fun. Flying is serious business. People die. Two Gemini astronauts, Elliot See and Charles Bassett, crashed in a T-38 in St. Louis in 1966. They flew up to check out the *Gemini 9* spaceship McDonnell Douglas was building for their upcoming flight. They got turned around in bad weather, couldn't find the runway, and crashed into a hangar—the same hangar that their Gemini capsule was in. Both astronauts were killed, and they took out their own spaceship with them. That same year, C. C. Williams died in a T-38 crash over the Everglades before he had the chance to fly in space, and Alan Bean replaced him on *Apollo 12*.

I wasn't worried so much about the flying. On that front, I felt good. The thing I was most scared about was that I didn't want to FOD the jet. FOD is foreign object debris. It's a big problem. The turbine blades on a fighter jet are razor-thin. Anything that's loose on the runway can get sucked into the engine and cause trouble: a soda can, a piece of glass. A pen that falls out of your pocket can get you killed. Every day the crew walks the runway and picks up anything they can find. It has to be pristine. Same thing in the cockpit. Anything that gets loose can jam the controls: A paper clip drops on the floor, you have to find it and remove it before that plane can move an inch.

The diligence and the mind-set you need to fly high-performance aircraft—or to fly in space—is totally different from the way you live in real life. There's no margin for error. At home I can be a bit of a klutz. So I was petrified I was going to FOD the jet. Frank and

I were scheduled to go out at 4:30 in the afternoon. As we suited up I kept peppering him with nervous questions, so many that we were running late. The last thing I said to him walking out was "Hey, Frank, I don't want to FOD the jet."

He gave me this strange look. I said it again and he said, "Oh. I thought you said you didn't want to *fly* the jet." He thought I was chickening out at the last minute. "Relax," he told me. "Everything will be fine.

I climbed up the ladder and into the cockpit. Bob Mullen followed me up and helped me with my parachute straps and my mask and my helmet and everything else, like a mother bird sending a baby chick out for the first time. Once I was strapped in I looked over at Bob for approval. He nodded, smiled, shook my hand, and said, "Have a good flight. See you in a bit." Then he climbed down and pulled back the ladder and I was on my own, about to go punch a hole in the sky. I did the radio calls to get runway clearance, and we taxied to line up on the runway, powered up the engines to make sure all was well, and then lit the afterburner. Frank released the brakes and we started accelerating, quickly. We were going over one hundred and fifty miles an hour when Frank raised the nose and we shot up into the sky. I felt like I was riding a rocket ship.

There are a couple things you have to do on your first flight, kind of like your initiation. The first thing is to go weightless. You fly up and push over and plummet straight down. Going weightless is an incredible head trip. I was strapped in tight, but I could still feel myself floating up a bit. My pen was attached by a lanyard to my kneeboard and it floated up for a moment, slowly, like magic. Dust floated off the dashboard, too. The weightlessness only lasted for a few seconds, but it left me with an unmistakable feeling: I wanted more.

The second thing you do is break the sound barrier. Frank flew us up to a high altitude again, because you get more speed flying down. Then you light the burner to get as much thrust as you can. The plane starts to shake and you're pinned to your seat and you watch the Mach meter inching up: 0.95, 0.96, 0.97 . . . When we reached 1.0, I said, "There's Mach 1," in my best Chuck Yeager impression, which was what I'd always dreamed of doing if this day ever came. There's a boom in the sky as you pass Mach 1, but you don't hear it. You don't hear anything, because it's behind you. You're moving too fast—faster than the speed of sound. The engines, the roar of the wind, it's all silent. The only noise is the sound of Air Traffic Control talking to you in your earpiece. And the view. *Wow.* Unlike in a commercial plane, with the T-38's clear canopy I could see all around me, a big blue sky spread out in every direction. It gave me the sensation that I was zooming and swooping through the air like a bird.

After going weightless and breaking Mach 1, Frank took me through some loops and high-g turns to work on my physiological training. Anytime you take a hard turn in the jet, the centripetal force will push the blood out of your head and down into your lower extremities. You'll become light-headed and possibly pass out—it's called a g-force–induced loss of consciousness, or GI-LOC. You have to grunt and tighten up the muscles in your body to constrict the blood flow so the blood stays in your head. It's the craziest feeling. You go into the turn and you can feel the light-headedness start. Then your peripheral vision starts to fail— you get tunnel vision. You're being pinned into your seat. Sweat's pouring down your forehead. You grunt and you tighten up and the blood pushes back up and your eyes come back. You find your equilibrium. You're training your body to endure the limits of what the human body is capable of.

I loved flying. I could not get enough of it. Backseaters had to log a minimum of twenty-five training hours in the T-38 every quarter. I was always near the top of my class in hours. I had more hours than any other mission specialist in my group, especially out of the civilians. Some of them looked at flying like a chore. To me it was the ultimate. I used to love putting on my flight suit and my aviators, hopping in my car, and heading over to the airfield. I'd blast rock and roll really loud on the way over, usually something like Smash Mouth or Bachman-Turner Overdrive, whatever was the loudest thing I could find on the radio. Granted, this was no longer the days of astronauts racing Corvette convertibles across the California desert; I was rocking out in our Nissan Quest mini-van. But I didn't care. It was still awesome. I'd park in one of the reserved spots marked AIRCREW and then stroll through the hangar past the shiny, clean T-38s. I'd stop in the flight planning room and meet my pilot; then we'd map out our flight plan and file it with Air Traffic Control. Then it was down to the parachute room to pick up our harnesses and out to the flight line to get in our plane.

The best part was that you could do it pretty much whenever you wanted. It wasn't like getting to space, where you were sitting around, waiting to be assigned. You could hop in a jet and go. The instructors were test pilots out of Pax River and Edwards Air Force Base. They loved having eager students because they loved to share the experience of flying, and they had the best stories, military exploits, launching off aircraft carriers, combat flights. Some of the older instructors had stories about showing legendary astronauts the ropes. These weren't the stories I'd read about in *Life* magazine. These were inside stories from the people who'd lived it, and I hung on every word. What amazed me was that they accepted me right away. I was a part of their military flying culture now. I belonged there. They were *Right Stuff* guys and we were flying together in *Right Stuff* planes doing *Right Stuff* stuff. They'd take me out and we'd do the craziest maneuvers: cloverleafs, aileron rolls, barrel rolls, Immelmann turns. It was unbelievable.

The single most fun thing to do was to go cloud surfing. When you're flying cross country in a commercial jet, you have to fly at the altitude allowed by Air Traffic Control. There might be clouds or there might not, and it's hard to tell how fast you're going without any physical points of reference. In the practice area you can do whatever you want. You find the cloud deck and dive down and nestle right in and glide along the surface, wisps of vapor whipping by your head, giving you the sensation that you're really moving.

The best was coming up on a big cumulus, a giant, puffy marshmallow cloud. You'd come up on the side of it and then roll into it and it's pure white all around. Sometimes the sun would break through and you'd see rainbows. And it's perfectly quiet. You keep the radio on low and only communicate when it's absolutely necessary as you soar through the sky. It's the closest thing to heaven you can experience on Earth.

We'd usually stay out until Bingo time. On some military jets, the fuel indicator has a warning sound that goes *bing-o, bing-o,* to let you know you've hit a certain fuel level. So you'd say "It's Bingo time" and head in. The coolest thing to do on landing was a touch-and-go. You'd make your approach and come in and tell the tower, "Touch-and-go, request closed pattern." They'd come back with "Touch-and-go, closed pattern approved." You'd touch down, wheels down, nose down, then *BOOM!* you'd jam the throttle and *WHOOSH!* you'd take off again. Then you'd get your speed up and make a tight turn at high speed and go back into your pattern. Then you'd go back and do it again. Then you'd go back and do it *again.* It was like riding the world's best roller coaster over and over without ever waiting in line.

In addition to spaceflight-readiness training, the added bonus was that we got to use the T-38s for transportation. NASA's operations are spread out all over the country: the Goddard Space Flight Center in Maryland; the Marshall Space Flight Center in Huntsville; the Jet Propulsion Laboratory in Pasadena; Ames Research Center in Silicon Valley. Astronauts don't fly commercial to those places if they don't have to. If you're a mission specialist who needs to visit Huntsville, you grab a pilot and you go. If you're a pilot who needs to go to Ames, you grab a backseater and you go. We get our training hours and we save the taxpayers the cost of an airline ticket.

Right after my first flight, Scorch came by my desk and said, "Come on. I gotta go to Yuma, Arizona. Let's get you some hours." That's how it worked. People were always heading out for one reason or another, and I purposefully tried to be the best backseater I could be so I'd be at the top of everybody's list to fly with. Flying these training jets cross country, you'd usually have to refuel. The wind might favor you going down to Cape Canaveral, but then you couldn't one-hop it home. You'd have to stop off someplace, and one of the best places was Acadiana Regional Airport in New Iberia, Louisiana. It was in the middle of nowhere and mostly serviced helicopters going out to oil rigs in the gulf. They had an FBO, a fixed-base operator, which had a contract to sell fuel to the government. It was run by a guy named Al Landry, and he catered to the military. He had a food concession at the airfield. Every day he had incredible Cajun food that he made right there. Crawfish étouffée, fried catfish, gumbo, jambalaya. He'd have tuna melts during Lent. Al was legendary. All the astronauts loved him. You'd go and sit around and eat while he fueled up your plane. Then you'd head home. This wasn't just allowed; it was *required*. You needed the twenty-five hours. You only got in trouble if you didn't fly enough.

Those first couple of months in the air were like a dream. I wasn't commuting to work on the Long Island Rail Road anymore. I was commuting to work in a high-performance jet. I knew flying was a dangerous, serious business, and I gave it the respect it deserved. But at the same time there was never a minute in that plane where I didn't feel like a kid, with that pure joy and exhilaration I used to get when I was playing make-believe in my backyard. Only it wasn't make believe. It was real. It was my job.

Carola and I were part of a parents group in the neighborhood, and every year they had a Christmas cookie exchange, something fun to do with the kids. That first December, one afternoon I ran over straight from Ellington after a flight, and I still had my flight suit on. I picked up Gabby and Daniel, we got our cookies, and went home. It was a beautiful sunny day in December, which in Houston is still warm, and I can remember lying out on the front lawn with my kids in my custom-made superhero flight suit, eating blondies and looking up at the sky and thinking: I was just there. I just flew. *I can fly.*

I wanted to grow up to be Spider-Man—and I did.

10

IF YOU HAVE A PROBLEM

When you watch *The Right Stuff* and see the training that the original Mercury Seven went through, you see those guys were basically treated like lab rats. They were being prepared to go to space, but nobody had ever been to space, so nobody actually knew what to prepare for. It was all a bit haphazard. That's not the case anymore. NASA's had forty years to work out the kinks. They know where you need to be, and they know exactly how to train you to get you there.

In addition to spaceflight-readiness training, one of the first things you do as an ASCAN is go on tours of the major NASA facilities. You meet everyone, the key players working across the country, but it's not just a meet-and-greet. There's a reason for it. Space is a daunting place. It can be terrifying, actually, and you need to know you're not alone up there. You need to know that every last NASA employee stands behind you. They also need to meet you so they can put a face to a name and know who they're protecting up there.

All of NASA's facilities are cool, but the coolest, hands down, is Kennedy Space Center at Cape Canaveral in Florida. It's mind-boggling when you see it up close. The Vehicle Assembly Building, where they built the Saturn V rockets and assembled the shuttle in our day, is the largest building in the world. There's a giant American flag down the side of the building. An eighteen-wheeler can fit inside each of the stripes on the flag. The interior of the building covers eight acres, and the volume of the space is 129,428,000 cubic feet—nearly four times the space inside the Empire State Building. The building has its own weather system. Clouds form. Birds nest up in the rafters. The crawler-transporter that takes the shuttle out to the launch site is the largest self-powered vehicle in the world. It's over three stories high and weighs nearly 6 million pounds. NASA takes the ASCANs out there and shows us these things so we get a good sense of the scale of the task we're about to undertake.

Back at the Johnson Space Center, for the first year, when you're not up in a T-38 or at the gym, you're probably in a classroom or a simulator. You learn everything about how the shuttle works, from top to bottom: the propulsion systems, the navigation systems, everything. Flying the T-38s, running simulations, studying the systems—month by month, piece by piece, they're building you up, transforming you into someone who's ready to walk out onto that launchpad at the Kennedy Space Center and strap yourself to the top of a bomb. It's a slow, deliberate, and very thorough process.

I went on the same tours and sat in the same classrooms as the rest of the Sardines, but my education ended up being different from theirs. It came much faster and was way more intense. Just months after joining NASA, I suffered what was then and may still be the most difficult ordeal I've ever gone through. It taught me an awful lot about my new job in a very short period of time. It was an experience that had absolutely nothing to do with space and everything to do with being an astronaut.

When I went to work at NASA my father was seventy-three years old. He'd been having serious health problems for a while. He was always a bit overweight, and he'd had a triple bypass eleven years before. In January 1997, about five months after I started, he had to have another bypass and get a valve replaced. When I flew up to see him, he looked terrible. His recovery wasn't going well. He'd developed a bad skin infection, and he was feeling worse and worse. Then his doctors started seeing problems with his blood work. They diagnosed him with myelodysplastic anemia, a condition where bone marrow fails to make the three types of blood cells: red blood cells, which carry oxygen; platelets, which help your blood clot; and white blood cells, which are a key part of the immune system. In the worst cases, myelodysplastic anemia can lead to acute myeloid leukemia, an aggressive type of cancer. Which was exactly what happened. In July the doctors told us my father had leukemia and he had, at the outside, six months to live.

I was in shock. I couldn't believe it. You get only a few people in life who love you truly and completely and unconditionally, people who will be there for you under any circumstances. My father was one of those people, and I couldn't imagine my life without him in it. I started flying up to New York as often as I could. Sometimes I'd fly commercial and take Gabby and Daniel with me. Sometimes I'd catch a ride on a T-38 if someone was going that way. I started going with him to his appointments. My dad was being treated at Sloan Kettering, a world-famous cancer hospital, supposedly the best of the best. But his doctors had basically thrown up their hands and said, "There's nothing we can do." They said his skin rash and his heart problems made him ineligible for the intensive chemotherapy this type of leukemia required. I asked about experimental treatments, new drug studies, anything. He was ineligible for those, too. They refused to treat him. But my father wasn't ready to give up, and I wasn't ready to let him go.

My father had it rough growing up. He grew up in the Great Depression, worked on a farm. That old line about walking twenty miles a day to school in the snow, that was actually his life as a kid. My dad was smart and popular. He was one of the smartest kids in his class. He went out for the football team, was voted class president. But he never got the opportunity to follow his dreams. He could never get too involved after school. My grandfather wouldn't let him do anything because he had to come home and work the farm. College wasn't even an option.

My father turned eighteen in 1941. His buddies were enlisting to fight in World War II. He wanted to be a fighter pilot, but the Army wouldn't take him. With five sisters and an ailing father, he was the only able-bodied male on the farm, and they needed food for the war effort. They deferred him. While his friends went off to Europe and the Pacific to defend the country, he stayed behind, growing carrots and onions to feed them at the front.

After the war, my grandmother died and the farm was sold. My father was able to move down to the city and get his job with the fire department. I know he loved the people he worked with and took pride in what he did, but I don't know if it's what he would have chosen to do if he'd been able to go to college and follow his dreams. He did what he had to do for his family. And because my dad never got to do anything, he made sure I got to do everything. I was going to get to live out every dream he never got to have. I wanted to go to the moon? Great. I wanted to pitch for the Mets? Fantastic. My dad played catch with me every night in the front yard. He came to all of my baseball games. All of them. Dads didn't do that back then. When I was growing up, dads were at work. Together, he and my mom gave me the room to dream as big as I wanted. I would never have been an astronaut if they hadn't done that.

I wanted my dad to see Gabby and Daniel grow up. I wanted him to see me fly in space. I wanted him to see my shuttle blast off into the sky and know that he'd been a great father, that everything he'd done for me had paid off in spades. I knew I had to do something, but I wasn't sure what. I'd hit a wall with the doctors at Sloan Kettering. I got frustrated with them for refusing to treat my dad. I was like "You can't treat cancer at a cancer hospital? Where are we supposed to go, the Dairy Queen?" So I said "Adios!" to those guys and turned to the best people I knew for help: my fellow astronauts.

Quite a few astronauts are MDs. Story Musgrave, Dan Barry, Lee Morin. I went to them and asked, essentially, "Hey, my dad's got this thing. I don't know anything about it. What can I do?" They did some research and said it didn't look good. They brought in the flight surgeons to talk about it, and one of them, Smith Johnston, recommended me to Dr. Elihu Estey, a colleague of his at MD Anderson Cancer Center in Houston who specialized in MDA and leukemia. Smith arranged for my father's records to be sent over and I talked to Estey on the phone. He said, "We'd be happy to try to treat your father," and he laid out the options of what they could do. It was an experimental protocol and a long shot, but if my father was willing to do it, they would give him a shot.

I said, "But what about his rash and his heart problem?"

Estey said, "Yes, your father does have a bad heart. That's not good. He does have a skin infection. That's not good, either. But you know what else your father has?"

"What?"

"He has leukemia. If you don't do anything he's going to die."

That October my mom and dad rented an apartment in Houston near MD Anderson and flew down to start treatment. The difficult thing about his treatment was that he was going to require many blood transfusions. You can get those from a blood bank, but it's better to have donors. Again I turned to the people I knew, the astronauts. One morning at work I wrote an e-mail explaining my father's situation, his treatment, his blood type, and what he needed, saying if anybody could help out and donate blood, I'd really appreciate it. I cc'd everyone in the office and sent off the e-mail and went back to whatever I was doing.

My computer made this *bing!* sound every time I got an e-mail. Maybe thirty seconds after I sent the e-mail I heard it. When I checked my in-box there was the first reply saying yes. For the rest of the day my computer went crazy. *Bing! Bing! Bing! Bing!* Yes. Yes. Yes. Yes. I'm almost certain every single person in the office who shared my father's blood type volunteered. And the people who weren't a match came by to say they were sorry they couldn't help and was there anything else they could do?

Then my father needed platelets. You have to donate a lot of blood to get very few platelets. That didn't matter. I sent out another e-mail and the same thing happened. *Bing! Bing! Bing! Bing!* Yes. Yes. Yes. Yes. Then, a few weeks after that, my dad needed white blood cells. Donating white blood cells can be risky. You're effectively lowering your own immune system by doing it. If your white blood cell count is off, it can affect your flight status; the doctors won't clear you. Didn't matter. *Bing! Bing! Bing! Bing!* Yes. Yes. Yes. Yes. Everyone offered to help.

I was floored, stunned, speechless. At that point I'd only been an astronaut for a year and I was still learning what the job was about. I knew that teamwork and camaraderie were an important part of it, but I didn't understand what that really meant until my father got sick. What it means is that if you have a problem, we all have a problem. If your father is sick, our father is sick. It's not just a mentality that exists to help each other get through flight simulations and survival training. It's something that encompasses the way you live your whole life.

All I'd done was go to Smith Johnston to ask for advice, and that turned into my dad getting the finest treatment available at MD Anderson and the entire NASA astronaut corps rallying behind me to get my family through it. Smith checked in with me once a week and kept up with my dad's records to make sure he was doing okay. I had no idea that would happen.

My buddy Scorch was the first person to volunteer to donate blood. He was like that with everyone. Somebody needed something, his hand would go up. He donated platelets for my dad, and just looking at a guy like Scorch, this huge, ripped Marine, you knew his platelets had to be in phenomenal shape. I'm no doctor, but I swear the day we gave that transfusion from Scorch to my dad, his blood count improved dramatically, like a miracle. That's when my dad started to turn the corner.

The astronauts' spouses volunteered, too. I was completely swamped, spending nights at the hospital, still doing my astronaut training, but somehow everything got done. The grocery shopping, babysitting for the kids—someone was always right there to help. Before my dad got sick, Kevin Kregel had asked me to be a family escort for the flight he was commanding that year, STS-87, which was going to go up in November. Being a family escort is an important job. Launches are especially tough on spouses and kids. In the weeks leading up to the launch, the escort's role is to be there for anything the family needs. During the launch you stay with the family every step of the way. Most important of all, you're there by their side in case their loved one doesn't make it back home. The launch came right in the middle of my father's treatment, and as I was flying back and forth to the Cape to serve as a family escort, everyone else in the office was serving as *my* family escort back home.

Rick Husband, an Air Force pilot from Amarillo, still waiting for his first flight assignment, was the other escort serving with me for STS-87. Rick was a pure soul, a good father and husband, didn't have a mean bone in him. I don't think I ever heard him curse. He was very religious, very Christian. He sang at his church, was into Christian music. He quoted scripture and read the Bible to his kids. He even videotaped Bible lessons for his kids to watch while he was in space. Escorting for that flight was especially difficult because of my father's illness. But Rick was someone you could talk to about anything and who you knew would do anything for you, no questions asked. We spent hours together going back and forth to the Cape. We'd have long talks, about our fathers, about life, about death. One day the hospital was trying to reach me. My father was having a particularly rough time and I was working, nowhere near a phone. The doctors called Carola and she tracked down Rick and he came and found me. We talked, and he asked me, "Do you want to pray about it?" He sat there and took my hand and the words flowed out of him, asking God to help my father and me and my family. With some people, that might seem like overstepping a personal boundary. I wasn't as religious as Rick, but with him it didn't feel uncomfortable at all. That's just who he was.

Kregel pitched in with a few words from orbit, too. He sent me an e-mail from the shuttle: "How's your dad doing? I've been praying for him up here. I'm closer to the Big Guy, and there's less static, so I think it goes directly through." I thought that was pretty cool. It's the way the world should be. Somebody gets sick, you visit the hospital, bring a plate of food, help get their kids to school. At NASA, any crisis you went through you weren't alone.

If you've ever wondered what the right stuff is, that's what the right stuff is—the *real* right stuff. It's not about being crazy enough to strap yourself to the top of a bomb. That's actually the easy part. It's more about character, serving a purpose greater than yourself, putting the other guy first, and being able to do that every single day in every aspect of your life. People ask me all the time what it takes to become an astronaut. It's not about being the smartest or having the most college degrees. The real qualifications for being an astronaut are: Is this someone I'd trust with my life? Will this person help look after my family if I don't make it home?

After seeing how the office reacted to my father's illness, I understood better why I'd been selected and how the selection process works. I had the PhD and good experience with robotics, but every other finalist had qualifications as good as mine. What it came down to was that they decided I had the right attitude to be a part of their team. They liked what they saw of my personality. Whether it was on the baseball field or collaborating with my friends at MIT, I'd always been a team-oriented guy, and the only way to put a spaceship into orbit is if everyone's working together. Very few jerks have been to space.

Astronauts are exceptional in terms of what they've accomplished, but in terms of where we came from we're regular people. What we have in common is this shared goal, to serve the people and push the boundaries of knowledge for all mankind. And the best thing about joining this club is that it comes with a lifetime membership. John Young may have been the only Apollo legend still on the active roster, but all those guys—Neil Armstrong, Buzz Aldrin, Jim Lovell—they were still a part of NASA. Many of them still live in Houston and come by often for business or personal reasons. These guys are our heroes, but there's this connection that's passed down with our group. A good story one of my friends tells is that when he got back from space, Armstrong was visiting. Neil stopped to talk to him and said, "You know, I haven't been up there for a while. Tell me what it's like these days." My buddy was like *Are you kidding me? This is* Neil Armstrong, *and he's* asking *me* about *my trip to space.*

That shared connection, that bond, is especially strong among the astronauts, but it extends to everyone at NASA. They took on my problem as their own and pulled my father back from the brink. He'd started this long-shot, experimental treatment in late October. In March the doctors looked at his blood levels and they were 100 percent normal. He was in remission. He was cancer-free.

For the first time in a year my family was able to take a breath and relax a bit. My dad was able to get around. I took him to the rodeo with the kids. Then, in April, the Sardines were ready to graduate from ASCAN to full astronaut. There was a ceremony up at the Johnson Space Center. Everybody had to dress like a grown-up: suit and tie, the whole bit. We were awarded certificates and each of us got our silver astronaut pin.

Because my father was in town and doing better, he was able to go to my graduation. By that point he was a bit of a celebrity around the astronaut office. Everyone wanted to say hello and shake his hand and congratulate him on pulling through. He met everyone who'd donated blood and platelets and thanked them for saving his life. After the ceremony we went out for this dinner with Charlie Camarda and his big New York Italian family from Queens. The service was terrible and our mothers complained so much that we got a ton of free food. It was horribly embarrassing, but I was happy that dad was finally back in good health and could be there to see me graduate. I wanted him to see that those early mornings and late nights taking the bus to work had paid off, and I wanted him to know how grateful I was for everything he'd sacrificed for me.

At the beginning of May, my father was doing well enough that my parents left Houston, moved back to New York, and started settling back into their old lives. He'd cheated death. Then, in late July, my father went in for a checkup with his New York doctors, and everything fell apart. It fell apart fast. Whatever the experimental protocol had done, the effects were temporary. The leukemia was back, full blown, and his condition was worsening by the day. He decided to come back to Houston to try treatment again. I'm sure he knew it wasn't going to work; he said he wanted the hospital to keep trying so the results might be used to help other people. He flew down at the beginning of August.

The day my father died, I was asked to be John Glenn's family escort for his return to space on STS-95. Glenn was seventy-seven years old, three years older than my father, and he was going to go back up for the first time since his original Mercury flight in 1962. It was going to be a big, high-profile deal. I couldn't wait to tell my dad about it. I drove down to MD Anderson that night after work. He wasn't in good shape. The problem with those chemo and radiation therapies is that the drugs that kill the cancer kill pretty much everything else, including the immune system, leaving you susceptible to infection. The doctors put him in this special sealed-off critical-care unit, but it didn't make much difference. He'd developed pneumonia and he had a terrible fever. His heart was giving him problems. He was dying. There was no more avoiding it.

We sat there in his room and we talked. I told him about the John Glenn flight. He smiled and he stared me right in the eyes and he said, "Do you know how proud I am of you?" I think he knew when he said it that we wouldn't be seeing each other again. We talked for a while longer, but he was tired and I let him get to sleep. I went home and was only there a short time when I got the call. I turned right around and drove back in, but I didn't make it in time. My father died on August 28, 1998.

The next day I was at home making arrangements to have my father's body flown back to New York for the funeral. Kevin Kregel called. He said, "I still need my flight hours this month. So I'm taking you home. You're in my backseat. I've already arranged it with Ellington. The jet's ready. You want to go now, we go now. You want to go later, we go later. You tell me when you're ready." At the funeral, my entire astronaut class sent flowers. The head of the Johnson Space Center sent flowers. Rick Husband sent flowers. Everyone knew what was happening and everyone wanted to help.

Getting selected to be an astronaut was amazing. Graduating from ASCAN and getting my silver astronaut pin, that was cool, too. But my father's illness and the way everyone in the office rallied around me—that was the thing that told me: *Mike, you're a part of this team. You're a part of this family, and what an incredible family to be a part of.*

A couple of months later, while I was serving as escort for John Glenn's family, his daughter Lynn told me an interesting story. John Glenn was so famous for being an astronaut that people forget he went up only one time in 1962. One flight, three orbits, four hours and fifty-five minutes. That's it. The story was that President Kennedy gave an order that Glenn was never allowed to fly in space again. He was a national hero. He was too valuable to lose if anything happened.

Glenn was being encouraged to use his stature to run for office, and after JFK was assassinated, Glenn left NASA to do just that: He ran for U.S. senator from Ohio. Alan Shepard, Gus Grissom, Gordon Cooper, Deke Slayton, Wally Schirra—they stayed and flew missions with Gemini and Apollo. They were part of building NASA into what it is today. Glenn didn't stay with the team. His star was too bright.

After years in the Senate, in 1984 he decided to make a run for president. Lynn was managing his primary campaign. It was a grueling experience; it didn't go well. One morning they were driving into the Senate Building, where he had his office. It was early. The streets were empty. She was following him in a separate car. All of a sudden he started making strange turns and going off in a different direction. She thought he was lost. She followed and caught up to him outside the Air and Space Museum, which was closed, but Glenn had parked, gone up to a window, and was looking inside where they had his Mercury capsule, *Friendship 7*, on display in the main gallery. He was standing there at the window, looking at his old spaceship. Then he turned to his daughter and said, "Sometimes I wonder if I made a mistake."

Here's a guy, he's world famous, he's a U.S. senator, he's running for president, but none of that gave him what he had when he was a part of this team: the feeling that he was a part of something special. When he went back up on STS-95, NASA justified it by saying it was to study the effects of weightlessness on aging or something or other. If you ask me, he just wanted to come back. He wanted to be with us. Leading up to his flight, he wanted to work out with us in the gym, tell some old war stories. The whole time he was so happy to be here. Being on this team, it's the same as anybody who's ever played on a Major League Baseball team. For the rest of your life, that's who you are. That's how you'll identify yourself. Long after you're gone, you still want to come back and hang in the dugout for a few innings and watch, because nothing in life compares to being a part of this.

11

SPACEWALKER

Other than flying in the T-38, one of my favorite parts of astronaut training was the enrichment lectures. Former astronauts and older NASA guys would come by and give talks about the space program. Chris Kraft, NASA's first flight director, came in to speak with us. So did Gene Kranz, the flight director played by Ed Harris in *Apollo 13*. My favorite lecturer was Alan Bean, who flew on *Apollo 12* and is one of the twelve guys who walked on the moon. After retiring from NASA, he became a painter. Alan's lecture was called "The Art of Space Exploration." He talked about the mistakes he'd made and how he learned to fix them. One lesson that took him a while to learn was that at a place like NASA you can only have an effect on certain things. You can't control who likes you. You can't control who gets assigned to flights or what NASA's budget is going to be next year. If you get caught up worrying about things you can't control, you'll drive yourself nuts. It's better to focus on the things right in front of you. Identify the places where you can have a positive impact. Concentrate there and let the rest take care of itself. The last thing Alan said to us was "What most people want in life is to do something great, and you have been given the opportunity to do something great. That doesn't happen often. Don't take it for granted. Don't be blasé about it. And don't blow it. A lot of times, believe it or not, people blow it."

Alan's lecture meant a great deal to me. I asked for a tape of it and watched it again from time to time. It got me thinking about where I could have a positive impact on the space program. As an ASCAN, you have to learn everything there is to know about how the shuttle works. Flying a spaceship is not like driving a car. If you're driving a car and something goes wrong, the worst that happens is you pull over and wait for a tow. You can't do that in space. As an astronaut, you have to understand everything on that machine that could possibly go wrong—anything from a broken toilet to a leaking fuel tank. Even though you can count on help from Mission Control, it's going to be on you and six other people to fix the problem, whatever it is. So during your training you learn it all. You get a copy of the *Shuttle Crew Operations Manual*, our textbook. You study handouts and workbooks for each of the individual systems: This is how the fuel tanks work. This is how the pumps work. You run simulations—"sims" in NASAspeak—to test your knowledge of how everything works, how to respond when systems fail. Sims are the backbone of your training as an astronaut. Because there's no margin for error in space, the solution to every foreseeable problem has to be worked out before leaving the ground. You run through a scenario until you find a fault in your plan. Then you address that and you run through it again. Then you run through it again. Then you run through it again and again and again until every possible outcome has been accounted for.

Once you've got the broad overview of the shuttle systems and operations down, that's when you're ready to graduate from

ASCAN to astronaut. From there, pilots take their training in one direction, learning how to fly the shuttle, and the mission specialists begin to specialize in a variety of particular jobs: spacewalking, flying the robot arm, serving as flight engineers. After my ASCAN graduation, I still wasn't sure where I'd end up. I'd wanted to be an astronaut my whole adult life. I'd wanted it so badly, I'd never really cared what kind of astronaut they'd let me be. The selection committee had chosen me not for any one particular skill but because they thought I had the general qualities of a good astronaut. I was still most all-around, I guess, same as in high school. For me, finding the place where I fit in was my biggest challenge. I had to find a specialty, my purpose.

Nick Patrick, my MIT buddy who became an astronaut in the class after me, once observed that the tragedy of being an engineer designing tools and instruments for space was that you never got to use them. Any engineer who's ever designed a car wants to drive it. Any engineer who designs an airplane wants to fly it. But because the odds of becoming an astronaut are so slim, the average aerospace engineer was doomed never to get to play with his own invention.

Luckily, I'd made myself the exception to that rule. I'd worked on the robot arm since my days at McDonnell Douglas, designing the video display that had flown on STS-69 while I was living in Atlanta. Its official name was the RMS Manipulator Positioning Display, and it had become a standard tool for robot-arm operators on shuttle missions. I thought it would be cool to keep working on that—to have the chance to drive my own car. In my DNA, I was a robotics and human factors guy. That's where I felt I had the most strength and, ultimately, that's where I felt I had the most to offer my fellow astronauts. That's where I would make my positive impact.

In addition to our training, every astronaut was tasked to work in a particular branch on a particular project for the shuttle or the space station. When you got assigned to a flight, you'd get rotated out to prepare for the mission. Then, once you returned, you'd be rotated back in to another assignment. Because of my experience I was assigned to the robotics branch. Around that time a new robot arm was being developed for the space station, the Canadarm2. I started logging a lot of hours of work on that. It seemed like the simplest, most direct path for me to get a flight assignment was to be a robot-arm operator on a station assembly flight.

Then I went to another enrichment lecture and I found the purpose I was looking for.

In the spring of 1997, Story Musgrave came to talk to us about spacewalking, or extravehicular activity—EVA in NASA-speak. Story Musgrave served as the lead spacewalker on the first repair mission for the Hubble Space Telescope and was probably NASA's most experienced spacewalker at the time. He was an amazing guy, an MD with a master's in biophysics and another master's in literature that he went and got after becoming an astronaut, just because.

Developing the ability to put a human being outside of a spaceship was one of those giant leaps that made space exploration possible. Without it we couldn't service the Hubble Space Telescope, assemble space stations, or walk on the moon. There wasn't a great deal of EVA done in the early shuttle era. For the most part, those missions were deploying satellites, conducting Spacelab experiments, things that didn't require working outside the shuttle. All that was about to change. More Hubble servicing missions were on the books, and those would involve extensive EVA. Space station assembly flights were about to get under way, too. Astronauts would be required to actually get out in space and put that whole thing together, module by module, piece by piece. We called it the Wall of EVA, and it was coming up soon.

Up to that point, rookies (first-time fliers) rarely spacewalked. For your first flight you got the contingency training and that was it. But with the amount of work that needed to be done on the station, the astronaut office realized that they would need rookies who could handle six- and seven-hour space walks on their first flights. A whole EVA skills program was being developed to train people for that. If you were a mission specialist interested in spacewalking and you showed an aptitude for it, you were likely going to get the chance.

At one point during Story's lecture, one person raised his hand and asked what you should do to be in the best shape, physically, to be good at spacewalking. Story said that weightlifting and building up stamina were all important, but the main thing you needed was to be big. "You need to be tall," he said, "with really long arms." The space suit is pressurized, which means there's resistance to every movement you make. The longer your arms, the more leverage you have; you're not fighting the suit. It's the same thing with your fingers in the gloves. Every time you make a fist or grasp the handle of a tool, you're working against resistance.

You also want someone with a long reach. Getting around in the suit is painstaking and exhausting. You want to be able to park in one place and be able to reach as far as possible without having to shift your whole body. Most important, if your partner is having a problem—if his or her suit has failed and they can't breathe—you have to be strong enough to get them back inside to safety. If you're a smaller person, with short arms and tiny fingers, you can go to the gym all day long, but that's only going to get you so far. "It's like the NBA," Musgrave said. "Every once in a while you get a little guy who can make a go of it, but generally the bigger you are the more of an advantage you have." I'm a big goon, six foot three. I have a long wingspan, hands that can palm a basketball. I sat there listening to Story and looked down at myself and I thought, *Here's something I was made for.*

My body was built to be a spacewalker. My mind, on the other hand, needed some work.

Most of the training for EVA takes place in the NBL, the Neutral Buoyancy Lab, which is a fancy name for an enormous swimming pool: 202 feet long, 102 feet wide, and 40 feet deep. In there we worked with full-scale mock-ups of the space shuttle, the Hubble, and the space station. You get in the suit and you're lowered into the water, attached to flotation devices that counterbalance the 200-pound suit you're wearing, making you neutrally buoyant and allowing you to move in an approximation of weightlessness.

As part of the ASCAN training, every astronaut had to get certified in the basics of EVA by doing four runs in the pool. Before we could get in the water in the space suit, we had to pass a high-grade scuba certification. I had a civilian scuba license, but this was a more difficult test. The hardest thing for me was the unassisted ditch and don. You had to swim down to fifteen feet, ditch your mask and flippers, go back to the surface, tread water, then dive back down, don the gear, clear your mask, and get back to the surface with no water inside your mask. I couldn't do it. I'd get halfway through donning my gear and I'd start to panic and need a breath and I'd have to shoot back to the surface.

Fortunately, I was at NASA, and my weakness wasn't seen as an opportunity to weed me out. It was a chance for the team to get behind me. Because helping my father battle leukemia wasn't enough, when Scorch saw I was having trouble he came by my office one morning, totally unsolicited, and said, "We're going to take care of that today." He found an astronaut with a backyard pool we could use and we drove over there. He sat in the pool with me and showed me what to do, showed me how to relax and get through it. All afternoon he dove with me to the bottom of that pool, up and down and up and down. We started in the shallow end and gradually moved down to the deep end. He took me through it over and over until I got it. A week later I passed the swim test.

When it was my turn to get in the pool wearing the suit, I was paired up with my buddy Charlie Camarda from Queens. Anytime you put me and Charlie together, we were like a couple of class clowns, like a two-man Three Stooges routine. We'd cut up and have fun. One of the things drilled into us during training was that spacewalkers had to stick close together on a run, which led to me and Charlie on the test standing next to the pool, dancing in our full-body propelyne underwear, and singing "Together Wherever We Go," an old Ethel Merman show tune from *Gypsy*.

The instructors loved it. Outside the pool Charlie and I were a hit. Inside, not so much. Your first run is what's called the Introduction Suit Qualification, a few hours in the water to get used to the suit and demonstrate your ability to maneuver in it. They lowered me and Charlie in, and it was a disaster. I tried to move around and felt like I was completely out of control. The suit is massive. I was stiff, clumsy. I felt like a Thanksgiving Day balloon, like the Michelin Man, the Pillsbury Doughboy. As I tried to translate down the side of the mock payload bay, inching my way along, I was thinking, *I'd better brush up on my robot-arm training, because this is not going to work.*

When ballplayers are having problems, their coaches always tell them, "It's all in your head." That's how it was with me and spacewalking. The human body can do amazing things, but only if the brain stops getting in the way. The first time you go in the water in a space suit, it's a totally disorienting experience. Everything is different. Your brain gets flooded with new and different stimuli. You get overwhelmed and start to panic. Your body is perfectly capable of doing it, but your head isn't doing a good job of telling your body what to do.

On a space walk, one astronaut is the free-floater and one astronaut is attached to the end of the robot arm. That way the arm operator can move them around to wherever they're needed. You attach yourself to the robot arm with these foot restraints by working your feet into toe loops and then spreading your heels to engage the locking mechanism. Your boot clicks into place. You can't spacewalk unless you can do it, and for the life of me I couldn't do it. When you're wearing the space suit, you can't see your feet; and since you're in the water, if you lean forward to try to see your feet, you pivot around your waist and your feet go up behind you. You have to do it blind. I could maneuver into the toe loops, but I could never get my heels into the locks. I'd try and try and try until I reached the point where I was so frustrated and overworked that I couldn't get anything right. A few times the divers had to come over and put my feet in the restraints for me so that the exercise could continue.

I ended up going to my neighbor Steve Smith, who had spacewalked on both of the last two Hubble servicing missions. "Steve," I said, "I don't know what I'm doing."

Like Scorch, Steve had total confidence in me. He said, "Don't worry about it. We're gonna take care of this. You're gonna be great at it."

Steve volunteered to do a run with me. We went out to the NBL an hour and a half early. We went over everything outside the pool. Then we got in the water and he showed me how to use the foot restraints. The problem was that I wasn't getting my foot flat. I was getting my toes in the toe loops, but I wasn't pushing my heel far enough down. Steve worked on that with me, too. He'd tell me, "Try it quick one time. Flat and go! Flat and go! And you're in." I practiced it over and over again until I could finally do it on autopilot. Steve showed me how to use the tools, how to translate up and down the payload bay. I learned to stay calm and in control of my movements. By showing me what to do, Steve gave me the confidence to not panic in the pool.

Becoming proficient inside that suit was a challenge—and that was a good thing. If there's one thing I learned about myself as an astronaut, it's that I need a challenge in order to do my best. If I already know how to do something or if it comes easy, I don't always give it my best effort. But if you tell me something is impossible—if you tell me I can't pass my MIT qualifying exam or that I'm medically disqualified from becoming an astronaut—then from that point forward, for whatever reason, I'm incapable of giving up. I cannot let that problem go until I know I've done everything in my power to try to solve it.

EVA training didn't come easily for me, but once I got the hang of it, I loved it. If putting on a NASA flight suit is like putting on a superhero costume, putting on a NASA space suit is even better than that. It's like having your own spaceship. Between the Hubble servicing missions and the assembly of the International Space Station, we were entering a new era when spacewalking would be at the forefront of everything NASA was trying to do. If they needed somebody in the pool, I was there. If I had a free afternoon, I'd go and scuba dive alongside other spacewalkers to observe their technique. If I couldn't do that, I'd volunteer to work outside the pool and support whoever was going in. I knew that this was where I'd make my positive impact, and I wanted to be ready when the opportunity came.

12

SHACKLETON MODE

In the history of human exploration, there are basically two types of people. On the one hand there are the scientists, men like Galileo Galilei. In seventeenth-century Italy, Galileo developed revolutionary telescopes, and with them he discovered the moons of Jupiter. He was the first person to identify the phases of Venus, proving Nicolaus Copernicus's theory of heliocentrism, that planets of our solar system revolve around the sun and not around the Earth. Scientists like Galileo work tirelessly in their laboratories, asking the big questions, expanding the limits of human knowledge.

Then there are the adventurers, guys like Ernest Shackleton. In 1914, Shackleton launched the third of his Antarctic voyages, the Imperial Trans-Antarctic Expedition—an attempt to cross the entire continent. His ship, the *Endurance*, was trapped and crushed in the Antarctic ice. For over a year, first camping on the ice and then taking rowboats out across the open sea, he kept his men alive and led them safely to rescue on South Georgia Island, off the coast of

Argentina. Men like Shackleton risked life and limb under punishing conditions to push the boundaries on the map, to expand our understanding of our world.

Some people dream of being Galileo. Other people dream of being Shackleton. The amazing thing about being an astronaut is that you get to be Galileo and Shackleton at the same time. You're tackling the big questions of human existence, and you're doing it in places where human life shouldn't even be possible. Down in the suburbs of Houston, driving around in our air-conditioned minivans, astronauts spend most of our time in Galileo mode, working on robot arms and other scientific endeavors. So, to learn how to survive in Shackleton mode, we have to leave the strip malls and the fast-food restaurants behind.

Starting with Skylab in the 1970s and then with Mir, the Russian space station launched in 1986, NASA worked to understand more about the effects of long-duration space flight on astronauts. We learned that people aren't machines; they're people. They get lonely, they get dehydrated, they're not sleeping right. You can only put them through so much before they start to break down. Nobody's ever gone crazy in space and turned into Jack Nicholson in *The Shining*, but some problems have occurred. The third Skylab crew kind of mutinied and quit working for a while.

When people are put in extreme circumstances for a long period of time, or even just removed from their normal routine, they get angry more quickly, teams split apart, trust and communication can break down. We called it "poor expedition behavior." With the launch of the International Space Station, long-duration spaceflights were going to be happening more and more frequently, and training astronauts to maintain good expedition behavior became a priority. We started hearing conversations about Shackleton around the astronaut office: How do we keep our crews together and functioning under impossible conditions the way he did?

As part of their contribution to the space station effort, the Canadian government offered to give astronauts cold-weather expedition training at its air force base in Cold Lake, a small town in northern Alberta that's home to a Royal Canadian Air Force base and weapons testing range. The Cold Lake expeditions started in 1999. Three of them were scheduled for the winter of 2000, one each in January, February, and March.

At the time I was still waiting to get assigned to a shuttle flight. Our astronaut class was huge, and the one before us wasn't small. There was a bit of a logjam. Then, making matters worse, thanks to forces beyond our control, the launch schedule slowed to a crawl. The assembly of the International Space Station had to be put on hold while we waited for a life support module being built by the Russians. It was a year behind schedule, and missions kept getting backed up on account of the delay.

I had no idea when I'd get flown. I was just working hard to improve in my EVA skills class, while keeping up with my day job in the robotics branch. Nancy Currie had taken over as head of the robotics branch the year before, and she showed me the value of having someone in my corner. She was a U.S. Army colonel with a PhD in industrial engineering who'd recently completed her third flight, flying the robot arm on STS-88 in December 1998. At that point the robotics branch was focused on training astronauts to operate the new robot arm being developed for the space station. Nancy picked me to help her with that, and we ended up working together closely. She was smart and had a great sense of humor. She got to see me up close, to see how well I worked with the arm and how well I got along with the training team. She became one of my biggest boosters around the office.

Around the end of 1999, Nancy started talking me up to Charlie Precourt, the head of the astronaut office. Precourt was an Air Force test pilot, highly decorated, fluent in French and Russian, an incredibly accomplished and, to me, intimidating sort of guy. They needed a robot-arm operator for one of the flights, and Nancy told me Precourt came to her and asked, "Who do you recommend?"

She said, "Massimino."

He said, "Well, we need someone more experienced."

She said, "Massimino."

She told him I had the hands-on experience and was the right choice. Even with her recommendation, I got passed over for someone else. Precourt didn't think I was ready. Then Cold Lake happened.

I wasn't set to go on any of the expeditions that year, but as the March trip was coming up, somebody had to drop out. Precourt was going to be personally leading the group, and he announced at a staff meeting that they were short one person and needed a replacement. Nancy Currie put me up for it. Then she came to me after this meeting and said, "I kind of volunteered you to go to Cold Lake."

I said, "Are you out of your mind?!" I knew I'd probably have to go Cold Lake at some point, but I wasn't jumping to volunteer. I'm still a kid from Long Island—natural outdoorsmen we're not.

She said, "No, it'll be good. You're going up there with Precourt."

I said, "Now I *know* you're out of your mind. You're gonna send me up to that frozen wasteland with the chief? With the boss? This is gonna be a disaster."

Precourt was the guy who could make or break you when it came to getting assigned to a flight. I was petrified of screwing up in front of him. Nancy said, "No, no. Look, this will be good exposure for you, good training. Charlie's a great guy, and he needs to know you're a great guy. This is going to be a good way for you to get on his radar."

She was right. I didn't want her to be right, but she was. Then she said, "Oh, and you need to get packed. You leave in two days."

Cold Lake was given the name Cold Lake for a reason: There's a big lake, and it's really, really cold. One night when we were there the temperature went down to minus 40 degrees Fahrenheit. But hey, it wasn't always that bad. Most nights it was a relatively balmy minus 20.

The expedition crew was Precourt and a bunch of new guys who hadn't flown in space yet, including me; Lee Morin, a former naval flight surgeon; Frank Caldeiro, who worked at the Kennedy Space Center as a propulsion system specialist before being selected to be an astronaut; and two of my astronaut buddies from MIT, Dan Tani and Greg Chamitoff. Greg had been selected by NASA two years after me, in 1998, and I wrote him a letter of recommendation—which I was glad to do on account of the fact that he saved me from my qualifying exam. Greg was also an Eagle Scout. He lived for this kind of outdoorsy survival adventure. I was glad to have him along.

We flew commercial into Edmonton and then bused another two hundred miles north to the base. There we met our instructors, these four hardcore Canadian Army special forces guys, the equivalent of our Green Berets. For them going camping on a frozen lake in the dead of winter was like going to the beach. Sergeant Colin Norris was the team leader, a total tough guy, big mustache. We stayed in the barracks for a couple of days and they took us through some basic training: how to build a fire in the snow, set up tents, tie knots.

They also set us up with our gear. They gave us these old two-piece long johns that looked like something your great-grandfather wore as a POW in World War I. We had thick wool socks, bulky coats, wool hats. We were issued Leatherman tools, navigation gear, and packs and that was it. But that was the point: to expose you to the elements as completely as possible while only giving you the bare necessities needed to survive. We were in full Shackleton mode.

After a couple of days it was time to head out into the field. Norris and his guys helicoptered us out to the middle of nowhere and left us. They would drop in on us once a day or so to check in and give us new instructions. They kept tabs on us, too. One night I was out on watch and it was pitch-black. I couldn't see a thing. The next day I found out that this Norris guy was about thirty feet away from me the whole time, watching me. I had no idea. If he'd wanted to kill me, he could have.

We landed late morning in a clearing in a desolate, icy, snow-covered wilderness. It was flat and endless, not a ton of trees but a lot of lakes, all of them frozen solid. We unloaded our packs and our sled and our gear. They timed us setting up our tent and trying to cook a meal for the first time. It was a disaster; we were totally clueless. Once we finally had everything set up, we watched the Canadians take off. As the helicopter flew away, leaving me and my five friends alone in the middle of the Canadian tundra, there was only one thought in my brain: *I'm cold.* I looked around and that's when it hit me. This is it, for ten days. I never realized how long ten days could be until I went to Cold Lake. For ten days I was never not cold.

We had to have the stove going inside the tent at all times;

otherwise it was too cold to do anything, which meant someone had to stay up on fire watch all night. We boiled snow for water to drink and heated up food in a pressure cooker. The tent didn't have a floor. We'd spread this foil sheet down under our sleeping bags and sleep on the foil on top of the snow. It was dark most of the time, too. Late sunrise, early sunset. We had only a few hours to get things done at zero degrees before it fell back down to minus twenty. We spent most of our time walking. As part of the exercise, every morning we'd break camp, trek to some new coordinates on the map, and set up camp again. Our gear would be loaded onto a sled, and we had to be our own sled dogs, hauling this thing across the frozen tundra. We had a GPS device but it wouldn't always work, because the batteries would freeze. I could use it for a couple of minutes, and then I had to hold it against my body for it to warm up again. The point of this was to induce stress. Over the course of a normal expedition it might take months for the stress to get to people. But we only had a couple of weeks in Cold Lake, so they had to accelerate the process.

This expedition was *rough*. Maybe we weren't stranded at the South Pole, but it was no joke. All kinds of things went wrong. One night the tent caved in on us and we had to tie it to a tree. One afternoon Lee Morin and I went out and got lost. We were so lost, we didn't even know we were lost; that's how lost we were. At one point we ended up following our own tracks, thinking they were taking us back to camp, but actually we were taking ourselves in a circle. We barely made it back to the others before dark.

Halfway through, Frank Caldeiro blew out his knee and had to be helicoptered out. I got water in my boot and ended up with frostbite, which gave me this burning sensation on the bottom of my foot. The pain was killing me through all of this trekking around, and it didn't completely go away until months after the expedition.

Cold Lake was so cold, I couldn't have any skin exposed outside the tent where the fire was. Ever. I forgot that one day when I was trying to tie a knot. I got frustrated with having big, clumsy gloves on. I pulled the gloves off to tie this knot and my skin was exposed for barely a second and it was like someone had taken an ice pick and jammed it through the middle of my palm, just a terrible, searing pain. At that point, I was like, *This is ridiculous. This sucks. I'm cold, I'm tired, I'm miserable. I want to go home and watch television and use a real toilet. Why are we even here? Why am I even doing this?* That was my own poor expedition behavior. Which meant the exercise was a total success. It pushed me to the point of having those feelings so I'd know how to recognize and cope with them.

One of the things our Canadian taskmasters would do was show up out of nowhere and give us spontaneous tasks. One time we had to move camp in the middle of the night: break everything down, load it, haul it, and set up somewhere else, in pitch blackness at twenty below. Another thing they'd do was drop food and supplies at random points and send us off to find it and bring it back to camp. Sergeant Norris came up to me with a map at dinnertime one night. He said, "Massimino, at 3:00 a.m. you and a team member are going to go to these coordinates across this lake and find a box of food and retrieve it."

I said, "I have to walk across that lake in the middle of the night?"

"Yeah."

"Are you sure it's frozen? I don't wanna fall in."

He looked at me. "You do realize that we're standing on a lake right now, right?" He stamped his foot. *Thump! Thump! Thump!* "Don't worry. You'll be fine."

I asked Chamitoff, the Eagle Scout, to go with me. Greg was having a ball; at one point he'd rigged up a shower for himself to bathe in the freezing cold. To me that was crazy, but I also knew Greg was the guy I wanted with me so I wouldn't lose a leg to frostbite or get eaten by a bear.

In the middle of the night we woke up and set off. We were halfway across the lake when we stopped and looked up. It was a perfectly clear night and the air was crisp and the stars were magnificent. There was no sound other than our breathing. Everything was perfectly still for miles around.

It hit me at that moment: I was having an extraordinary experience. I was out at the edge of civilization. Yeah, I was cold and, yeah, it was hard, but I was doing something amazing in spite of myself. I was learning new things about myself. I was being given the chance to step outside of my everyday life and look at the world in a completely different way. The world that had seemed so small growing up in Franklin Square was now vast and wide-open and filled with incredible, beautiful things. I turned to Greg and said, "Hey, remember how a few years ago we were a couple of kids in a dorm room, dreaming about becoming astronauts? And now here we are."

"Yeah," he said. That was the last either of us spoke. We just stood there, two buddies gazing out at the universe from the top of the world.

The whole trip changed for me halfway across that lake. The conditions hadn't changed, but my mind-set had, and that's what expedition training is for. I started to enjoy what I was doing. I started to appreciate the opportunities to learn new things, and the days flew by.

Survival training was not the point of the trip. We had food and water; they gave it to us. The goal was expedition training, learning how to deal with harsh, extreme circumstances. Shackleton's Antarctic expedition was a disaster. His ship was destroyed. He never reached the South Pole. Yet today he's revered because he kept his men together through such a catastrophic situation. He kept them focused on what needed to be done. He kept their minds active. He kept morale up. Shackleton was a great leader, and in any remote, difficult situation, leadership is key.

In Cold Lake, during our different tasks and exercises, we took turns leading the group. Before going up there, I was never comfortable in leadership roles. Now, for two out of the ten days of our expedition, I had to be the guy in charge—which was especially awkward because I was far and away the least experienced guy out there. Chamitoff was the Eagle Scout. Morin was older than me and a highly decorated officer who'd served in the Gulf War. Precourt was my boss. But I had to give the orders. I wasn't comfortable with it, so I handled it the way I try to handle most things: by telling jokes. The whole day I kept trying to get everyone to laugh so they'd be distracted from the fact that I was in way over my head.

At the end of each day the Canadians would grade us on how we did. What Sergeant Norris told me was something I never would have thought of: Humor is a great leadership tool. Most leaders, even if they're naturally funny, they'll get serious in front of the group and try to motivate people either by inspiring them or by cracking the whip. But if you can keep people laughing while they're freezing their butts off, that's good, too. My team completed its tasks and ended the day ready to go back out and do it again, which meant I'd done a good job. Precourt even took me aside and said he thought I'd done well. Nancy Currie was right. I was nervous about going out with the head of the office, worried that he'd see me screwing up. And he did see me screw up, but he also saw me work hard and get better. He saw what my strengths were.

When we first landed in Canada, Precourt and I went to exchange some money. I didn't know how heavily the exchange rate was in our favor. I handed the lady at the counter $50 and she gave me back $70. I said, "Hey, this is better than going to the race-track." Precourt thought that was hilarious, and from that moment on we were cracking each other up. I got to see that he was a regular guy. He'd gone to the Air Force Academy and learned all these languages and flew F-16s, but he was also a kid from Boston who loved hockey. We'd have picnic lunches and sit around shooting the breeze in the freezing cold. We both loved *The Godfather* and we cracked each other up quoting it the whole time.

At one point the two of us were lashed to a sled, with Precourt right behind me, pulling a couple hundred pounds of gear together. That sort of thing gives you a bond you can't get anywhere else. You don't get it playing golf together or even going to a ball game. Before we left, Precourt had been up on this unapproachable pedestal. When we got back to the office, he was definitely still the boss, but we were colleagues, friends. I'd see him in the hallway, talking to some of the other higher-ups, and he'd say, "Hey, Mass, I was just telling these guys about that time in Cold Lake. . . ."

About a month after my Canada trip, Nancy Currie and I went to Japan to work with a team of engineers with the Japan Aerospace Exploration Agency (JAXA) at their facility in Tsukuba Science City, a couple hours outside of Tokyo. The Japanese were working on a different robot arm, one to help conduct experiments outside the station and bring them into an airlock. It was incredible to go to a place like Japan and be there as an astronaut and not as a tourist. For American astronauts to visit the Japanese space agency was a big deal. We were treated like big shots, VIPs. They wanted to hang out with us. Neil Armstrong and John Glenn still cast a long shadow. When it came to the work on the robot arm, my opinion was very, very important to these engineers. Every point I made or problem I pointed out, they were hanging on every word. It was another reminder of how much power and responsibility comes with the position.

The experience was similar to when I first became an astronaut and people started treating me with deference and respect because of my job title. Back then I didn't feel like I'd earned it. This time it was different. This time I *felt* like an astronaut. I had been working and training for almost four years. My knowledge of the space program had increased a hundredfold. I was a part of a team of extraordinary people. I'd overcome challenges and obstacles and proven to myself that I was capable of doing things I never would have thought I could do. It's a funny thing. In the beginning, I felt like an imposter telling people I was an astronaut because I hadn't even been to space. Then I eventually realized that I was thinking about it all wrong. Going to space doesn't make you an astronaut. Being an astronaut means you're ready to go to space.

part

4

The Door to Space

13

SEEING BEYOND THE STARS

Imagine you're standing on top of the Empire State Building in Manhattan holding a laser pointer. Now imagine I'm down in DC on top of the Washington Monument holding up a dime, and you're able to hit that dime with your laser pointer. Now imagine that you and the Empire State Building are moving 17,500 miles per hour in one direction, and the Washington Monument and my dime are moving thousands of miles per hour away from you in a different direction, and you can still hold that spot on the dime even as we hurtle away from each other in opposite directions at incredible speeds.

That's what the Hubble Space Telescope does. That's how amazing it is. The Hubble stands right up there with the Pyramids and the Great Wall of China as one of the great engineering triumphs in human history. It's named for Edwin P. Hubble, the astronomer who discovered that galaxies like ours exist outside the Milky Way and first established that the universe is expanding—the scientific breakthrough that led to the big bang theory. Scientists were theorizing about the advantages of putting a telescope in space almost as soon as we started building rockets. A space-based telescope would be able to observe light undistorted by the turbulence in Earth's atmosphere. It would also be able to observe ultraviolet and infrared light, both of which get absorbed by Earth's atmosphere. A space-based telescope would be able to see things and learn things that no human had ever dreamed of.

The Hubble does all of that and more. It takes thermal images of faraway planets, helping determine which ones might be capable of supporting carbon-based life. It measures the distance between stars with incredible accuracy. It's shown us how fast the universe is expanding and exactly how old it is (13.8 billion years, in case you were wondering). Hubble discovered Pluto's four new moons. It helped us learn about how stars are born and how black holes are formed. Much of what the telescope has discovered are answers to questions we didn't even know how to ask. The Hubble is, without question, the single most important tool humankind currently has for understanding the universe and our place in it.

How the telescope works is nearly as incredible as what it does. There's no propulsion on the telescope itself. It collects energy through solar arrays and uses that to power internal reaction wheels. The wheels spin, and the shifting mass of the spinning wheels points the telescope. There are six gyros that spin as well, keeping the telescope fixed on its target as it hurtles through space. The telescope itself is housed inside a spacecraft brilliantly engineered to protect it. As it orbits the Earth, going from day to night, the temperature on its outer surface swings from 200 degrees Fahrenheit to –200 degrees Fahrenheit and back again. Even under those punishing conditions, the inside of the telescope is controlled to a comfortable room temperature, keeping the instruments perfectly calibrated to perform their tasks.

When the Hubble launched, it was the biggest story in space exploration since the launch of the shuttle itself. It had been in development since the early 1970s and was supposed to be deployed in 1983; technical delays kept pushing it back. In December 1985, I was still working at IBM, rushing through Penn Station to catch my train, when I glanced at the newsstand and the cover of *Life* magazine caught my eye. It showed a spacewalking astronaut pulling open a tear in the fabric of space to reveal a distant yellow-orange nebula. The headline read: SEEING BEYOND THE STARS: A PREVIEW OF AMERICA'S BIGGEST YEAR IN SPACE. I grabbed a copy and read it right away on the train.

Because of the *Challenger* disaster, 1986 turned out to be a big year in space for all the wrong reasons. The shuttle was grounded, and the telescope's deployment was pushed back again, to 1990. By the time it finally launched, what was supposed to be a $575 million project had ballooned into a nearly $1.8 billion project—and only after it launched did we discover it didn't work. The telescope's eight-foot-diameter mirror was defective, having been ground incorrectly. At its perimeter it was too flat by approximately 2.2 micrometers, which is a minuscule variation—less than the width of a human hair—but it was enough that the telescope was bending the light wrong and wasn't able to focus. The imagery from the visible light spectrum, those cool pictures of faraway galaxies and nebulae that everybody wants for wall calendars and screen savers—the telescope wasn't getting any of that. It was a disaster, a huge embarrassment for the space program.

Fortunately, we had the opportunity to fix it. Sophisticated, high-performance machines are temperamental. They need tender loving care to keep working right, and the Hubble is no different. That's why it was built to be serviced by astronauts. When it was launched, NASA had budgeted for four servicing missions that would go up and make repairs and upgrade the telescope's equipment as newer technology became available. After the problems with the mirror were discovered, the first of those servicing missions became a rescue mission.

The mirror itself couldn't be replaced. At eight feet in diameter, it's too big, and it isn't modular like some of the other components of the telescope. But even though the mirror had been ground incorrectly, it had been ground incorrectly with such precision that we knew *exactly* how much it needed to be adjusted, and we were able to fit it with corrective lenses: We gave it glasses, essentially. The first Hubble serving mission, STS-61, was launched in 1993. That crew installed the COSTAR, or Corrective Optics Space Telescope Axial Replacement, a device that put coin-sized mirrors into the light path to bring the telescope's imagery into perfect focus. STS-61 was probably, up to that point, the most important shuttle flight in the history of the program.

The second servicing mission flew in February 1997. As intended, that mission was mostly routine maintenance, a 30-million-mile checkup. They replaced some worn-out equipment and installed two major new instruments: the Space Telescope Imaging Spectrograph (STIS) and the Near Infrared Camera and Multi-Object Spectrometer (NICMOS). The third mission was originally supposed to be routine maintenance and upgrades as well, the 60-million-mile checkup. Then, in 1998, Hubble's gyroscopes started to fail, one after another, much sooner than expected. Fine, hairlike electrical wires suspended in the fluid inside the gyros were corroding, something the engineers never anticipated. Of the six gyros on board, at least three needed to be working in order for the telescope to perform its functions. As they continued to fail, we put the Hubble into an emergency low-power mode, like putting a laptop to sleep. It was alive and it wasn't falling out of the sky, but it wasn't taking any pictures or doing any science.

The third servicing mission became an emergency flight to replace all six gyroscopes. Some of the scheduled upgrades had to be postponed while the telescope was rescued—again. We needed to add an extra servicing mission but didn't have enough funding for five missions on the books. So Servicing Mission 3 became Servicing Mission 3A and Servicing Mission 3B, which was a bit of budgetary sleight of hand that allowed us to get everything approved. Servicing Mission 3A launched right before Christmas in 1999. That crew got Hubble working again. Now we had to put together Servicing Mission 3B to handle the remaining repairs that 3A didn't complete, which meant that a whole new flight and a whole new crew had to be assembled from the ground up on short notice.

From the telescope's deployment through all three servicing missions, no rookie had ever spacewalked on Hubble. The work was incredibly complex and the stakes were considered too high. If something went wrong on a station assembly flight, we always had the opportunity to go back and fix it. With Hubble there was no margin for error. Now, though, with so much EVA needed for the station, it was decided that the fourth spacewalking position on the team might go to someone new. The chance to spacewalk on Hubble was probably the most coveted assignment in the entire astronaut office. I wanted it the same as everyone, but I didn't imagine I'd be in the running for it.

When I came back from Japan, I was transferred from robotics to the EVA branch, which was being run by John Grunsfeld, who'd just come back from spacewalking on Hubble Servicing Mission

3A. Grunsfeld is the smartest guy in the room, even at a place like NASA, where everybody's already smart to begin with. MIT undergrad, University of Chicago PhD. He's not a big guy, but he's good in the suit nonetheless. He's an astronomer, and he loves the Hubble Space Telescope as much as anyone has ever loved anything. He's also mechanically inclined, which made him a perfect fit for the Hubble servicing missions. Even with all the astronauts who'd worked on Hubble in the past, Grunsfeld was well-known as the go-to expert on the telescope around the office. Shortly after I was transferred, Grunsfeld came up to me in the office one day and told me that he and Precourt wanted me to work on the development runs for Servicing Mission 3B.

Development runs are different from training runs. Training runs are to train the astronaut. Development runs are more for the engineers. Whenever they devise a new tool or a new method for tackling an issue—they've built a new ammonia tank, for example, or they need to repair and replace a cooling system—they need to try it out, test the device, test their hypothesis. The astronaut's job is to get in the water and help them work out the new procedure. This works, this doesn't. This is good, this is bad. Getting assigned to those development runs was the first step to getting assigned to the flight. There were other, far more experienced spacewalkers available to do them, but after Cold Lake, Precourt told Grunsfeld he wanted me. And I got the nod.

I knew it was a huge opportunity, but I had no idea how huge until the day we started. The briefings for the development runs were being held in the planning area outside the pool at the Neutral Buoyancy Lab. Every other time I'd been called in for a development run, we had maybe a couple of mid-level engineers briefing us on whatever project they were doing. But the moment I set foot in that first Hubble development run, I knew I wasn't in Kansas anymore. It wasn't like any briefing or any meeting or any sim I had ever been to at NASA. They had the full-size mock-up of the telescope and different instruments and tools and machines laid out. There had to be at least thirty, forty people there. The veteran astronauts who'd flown on the first three servicing missions—several of them were on hand. This was the A-team.

The Hubble was built by Lockheed Martin in California, and several of its instruments were built by Ball Aerospace & Technologies Corp. in Boulder, Colorado. (It's part of the same company that makes the Ball mason jars you have in your kitchen.) All their top Hubble people were there. The Hubble team from the Goddard Space Flight Center was there, too. Frank Cepollina, whom everyone called Cepi, ran the show at Goddard. He was well past retirement age, but he loved what he did, so he kept doing it. Cepi looked like one of my old Italian uncles: balding, always chewing a stick of gum. I think he did that to keep his jaw loose, as he also talked a lot, mainly about what needed to be done, how this mission was important, and why this telescope was the most valuable scientific instrument in the world.

I called Cepi the godfather of the Hubble. He was a visionary, a tough, energetic, workaholic type of guy. Goddard runs all kinds of science and weather satellites, and when Hubble came along, Cepi was the person who first saw the benefits of making a space telescope serviceable by astronauts. Normally it would have made sense for the Hubble operations to be run out of Houston, since that's where the astronauts are, but Cepi had the expertise and the political clout to have the servicing piece of the Hubble headquartered at Goddard.

Like Cepi, the Hubble teams from Goddard and Ball and Lockheed Martin weren't a bunch of kids fresh out of college. These were people in their sixties and seventies, the men and women who'd built the telescope twenty years before, who'd been through the construction delays and the previous servicing missions. The Hubble was their life's work. Some of the engineers who were there for that meeting had been brought out of retirement to come back and help. Ron Sheffield was Lockheed Martin's EVA and crew systems manager for Hubble. Back in the 1980s he'd led the team that built Hubble so that it could be serviceable, and he was still involved, teaching the astronauts how to perform every task, how to undo every connector, how to turn every bolt. He was a walking Hubble encyclopedia.

Going into that development run, it was like in the movie *Armageddon* when there's an asteroid the size of Texas headed straight for us and every top scientist on Earth has gathered to figure out a plan and the president is waiting on the line because it's that important—that's how it felt. There was an electricity in the room, this feeling that something big was about to happen. Everyone was happy, smiling, energized, ready to dive in. I remembered that in his enrichment lecture Alan Bean had told us that being an astronaut would one day present us with the opportunity to do something great. I knew as soon as I walked in that this was the moment he was talking about. Right in front of me was the opportunity to do something truly remarkable with my life.

It also felt like traveling in a time machine back to the Apollo days. With the moon shot, NASA's directive was to dream big and go for it. I quickly learned that the Hubble program was the same. Its budget was the envy of everyone else in the space program. Whatever you needed, you got. When it came to servicing and maintaining this piece of machinery, no expense was spared. But I could definitely tell, being in that room, that the extra money and attention and manpower came with added pressure: We could not screw this up. There was no margin for error.

We did these development runs through April and May. During one of the briefings we took a break, and I was standing under the mock-up of the telescope looking up at it. Grunsfeld was standing next to me. I said, "John, this mission is really important. I hope the run goes well."

He said, "I hope it goes well for you, too."

I said, "What do you mean?"

"You must realize you're on the bubble."

"What does that mean?"

"You're on the bubble of getting assigned. You're in the mix now. If these development runs go well, this is going to help."

I was speechless. I'd been hoping to get assigned to a space-flight, any spaceflight. But Hubble? That was too much to hope for. Grunsfeld was deeply involved in the program, and Steve Smith had flown multiple missions to the telescope. They were Hubble guys. That's what they were known for. Bob Curbeam, who flew on a couple of station assembly flights, used to say, "Hubble guys are the Jedi. The coolest." I wanted to be a Hubble guy. No rookie had ever done it, but there I was, standing there in front of the Neutral Buoyancy Lab mock-up of the telescope with the Hubble team around me and Grunsfeld putting it out there as a possibility. As crazy as the idea seemed in my head, Grunsfeld didn't seem to think it was crazy at all.

When I was selected to be an astronaut, any chance to go to space would have satisfied my life's ambition. But I have to say that once I was exposed to the Hubble mission, I wanted that flight more than any other. Everyone else wanted it, too. Every spacewalker at NASA had his eye on that Hubble flight, and many of them were proficient in the suit. I didn't start out as the strongest spacewalker in my class, and I was still working to catch up. What I think Pre-court and Grunsfeld both recognized was my determination. Every obstacle that was put in my way, every challenge that I faced, I doubled down and worked harder and figured it out and got past it. And that's the kind of person you need on a mission where failure is not an option.

From the moment Grunsfeld told me I was in the mix, that telescope became my world. For the next two months I did everything I could to learn everything there was to know. I watched tapes of the previous servicing missions. I sought out the guys who'd spacewalked on Hubble before and picked their brains for information. I went and talked to the engineers from Goddard who'd been working on the telescope for the past twenty years, absorbing all of their knowledge and experience. We'd get a briefing on Friday for a development run happening Monday, and I would go in on Saturday to walk through the steps on the mock-up to make sure I knew everything and was completely prepared before going in the water. Even if it wasn't my turn to go in the water, I cleared my schedule and went and observed.

The guy I leaned on the most was Steve Smith. In addition to helping me in the pool when I was learning to spacewalk, Steve was a good neighbor and friend and had been to Hubble twice. Since our daughters are the same age and we were neighbors, we were hanging out all the time, talking about spacewalking, talking about the telescope. I was constantly asking him questions, picking his brain.

Steve was also Charlie Precourt's deputy, so he knew a few days before everyone else who was going to get assigned to flights. That August, on the Friday before my birthday, Precourt let Steve know that I was going to get assigned to the next servicing mission—the first rookie ever to spacewalk on Hubble. Normally you find out you've been assigned when Precourt calls you personally, but Steve asked if he could be the one to tell me. Precourt said, "You can tell him, but you can't tell him until Monday."

"But it's his birthday on Saturday," Steve said. "I can't tell him on his birthday?"

"No."

The whole weekend went by. We spent Saturday together, taking our kids to Home Depot to get wood for school projects, and Steve said nothing. Then, on Monday morning, I was getting ready for work and Carola came in and said, "Steve's here." I thought, *What is Steve doing here at 7:30 in the morning?* I went to the front door to meet him. He handed me an illustrated children's book about the Hubble Space Telescope. I said, "What the heck is this for?"

"I think you better read up on this," he said, "because you're going to Hubble."

14

READY TO GO

Joining a shuttle crew takes you into an entirely different phase of being an astronaut. If NASA is like a team, a shuttle crew is like a family. My new family was the crew of STS-109, Hubble Servicing Mission 3B. We would be flying on the space shuttle *Columbia*, scheduled to launch in exactly eighteen months, in February 2002. Because this mission had so much EVA, the spacewalkers were assigned several months before the rest of the flight crew to give us extra time to prepare. In addition to me, the others were John Grunsfeld, Jim Newman, and Rick Linnehan.

When the teams were assigned, it went without saying that Grunsfeld would be EV1, meaning he would lead the space walks on days one, three, and five. He had the most experience, having spacewalked on Hubble before. Newman was named EV3. He would lead the space walks on days two and four. Linnehan had flown but never spacewalked, and I was the pure rookie. He was named EV2 and paired with Grunsfeld for three space walks. I was named EV4 and paired with Newman for the other two.

For me and Grunsfeld, STS-109 was the beginning of a strong friendship. He was a mentor who, over time, became a partner. Between the missions we flew and the work we did together in between, I've probably spent more hours in the company of John Grunsfeld than any other astronaut. Our personalities are very different, but in a good way. We balanced each other. I'm the loud, gregarious people person, and he's more on the soft-spoken side, a thinker, deliberate and capable. If we encountered a problem, he was the first one to come up with a solution, and most of the time he was right.

Grunsfeld stuck his neck out for me, too. After I was assigned to Hubble, he and I were out for a jog in the blistering Houston heat. He told me that during an EVA branch meeting there was some second-guessing about whether or not I was the right guy for the mission, given how new I was. Grunsfeld was the person who calmed everyone down and said I would do okay. I told him I wouldn't let him down.

Jim Newman I knew from working together on computers and robotics; he was the astronaut responsible for flying and testing my robot-arm display when I was at Georgia Tech. We'd worked together for almost ten years, and now we were going to be space-walking together. Newman had some idiosyncrasies. In a place like NASA, which is all about the team, he stood out as something of an individualist. His nickname was Pluto, not after the dog, but because he was in a whole different orbit. He's one of those guys who's so smart, he's often marching to his own drummer. Which is a good thing: You need somebody who tackles problems from a different angle. Newman was a good person to be partnered with because he was very experienced and saw it as his responsibility to bring me along and help me out.

Rick Linnehan was someone I didn't know that well before

being assigned. Rick had an interesting background for an astronaut: He was a large-animal veterinarian. He started out doing research at Johns Hopkins University and the Baltimore Zoo before
going on to do marine mammal research with the U.S. Navy. Very
funny, loved dancing to Johnny Cash and doing old Three Stooges
routines while floating in space.

A few months later the flight deck crew was announced. Scott
Altman—"Scooter"—became our commander. He and I were already old pals from five years of being neighbors and reenacting
scenes from *Top Gun* on T-38 flights. Nancy Currie was named our
flight engineer and robot-arm operator. She was and continued to
be an ally and advocate for me. She was a real veteran, too. Both
she and Newman had flown on the first station assembly flight, so
she brought some good experience in dealing with a high-profile,
high-pressure situation, which we were going to need.

Duane Carey—"Digger"—became our pilot. Digger was the
other rookie on the flight besides me. Like Linnehan, he was someone I didn't know that well going in. He was your classic Edwards
Air Force Base test pilot, loved riding motorcycles—a real *Right
Stuff* kind of guy. He looked the part, too, with the crew cut and
everything. The reason I didn't know him that well was because he
was never around much. If he wasn't working he was home with
his family, doing math homework with his kids. But once we were
assigned together we bonded quickly as the only rookies.

On a shuttle flight, the commander is in charge of the crew.
What he says goes. This was Scooter's first commander slot, but he
took to it naturally. Every shuttle crew is assigned its own office for
the months leading up to the mission, and our first task as a team
was moving in and setting up. It was such a small thing, but even
there I could see how the team dynamic was going to work. Scooter

walked into the office on day one and looked at what we were doing and said, "No, this needs to go here, that should go there. . . ." He was calm, confident, and in charge. He fell right into his leadership role, and we fell in right behind him. That, to me, was a good sign.

One of the first things we did as a crew was go out to dinner together. I could see the family dynamic already starting to form. Scooter was the dad, and I was the little brother. Linnehan and Digger were like the siblings I would goof around and have fun with. Nancy and Grunsfeld and Newman were like the older siblings I'd go to for advice. I was the rookie spacewalker and the only crew member still in his thirties. I was always in the position of asking questions and wanting to learn, and the others had this natural instinct to want to help me and support me.

After 109 was over, Charlie Precourt told me why he chose me. I was good in the pool and proficient in the suit, but so were a lot of people. The thing that set me apart, to him, was my personality. No matter how stressful the situation, I try to keep things light and fun, like I'd done up in Cold Lake, like I'd done going back to my high school sports teams. I was always the glue. This was going to be a difficult mission in a high-pressure situation. There were all these very different, very strong personalities sitting around the family dinner table, and having a fun little brother sitting down at the end broke the tension and balanced everything out. I didn't know that was my job at the time, but in hindsight it made perfect sense.

Once the team was assembled, we threw ourselves into the mission. We had a full slate of tasks ahead of us. We would be replacing the telescope's solar arrays with newer, more efficient ones; replacing the PCU—the power control unit, the central nervous system of the entire telescope; swapping out the older Faint Object Camera (FOC) for the exponentially more powerful ACS, the Advanced Camera for Surveys; and installing a new cryocooling system for the NICMOS, the Near Infrared Camera and Multi-Object Spectrometer, which had been dormant since its original cooling unit failed in January 1999, two years after being installed.

These were complicated, delicate upgrades to perform, and we only had eighteen months to prepare. Based on mission priorities and the amount of time each task was going to take, it was decided that on EVA #1, Grunsfeld and Linnehan would replace the starboard solar array. On EVA #2, Newman and I would replace the port solar array. EVA #3 would be the PCU replacement. EVA #4 would involve swapping the FOC for the Advanced Camera for Surveys. And on EVA #5, Grunsfeld and Linnehan would install the NICMOS cryocooling system. In addition to those main tasks, each day we'd also have a number of smaller, routine chores, like adding new insulation blankets and helping prep for the next day's space walk.

In terms of advancing the Hubble's capabilities and its scientific mission, installing the Advanced Camera for Surveys was the most important aspect of the mission. It was going to improve the Hubble's ability to capture images by a factor of ten. It was going to peer farther and deeper into space than any other instrument ever created. But from the point of view of preparing for the mission, swapping the ACS for the older instrument was the relatively simpler task. It wasn't *easy*. Nothing in space is easy, but the job itself was straightforward. Both the FOC and the Advanced Camera for Surveys are about the size of a refrigerator; each one is a big metal box. Because the engineers at Goddard and Lockheed designed the Hubble to be serviced, all we had to do was demate the connectors, disengage the latches holding the FOC in place, pop it out, slide the ACS in, and hook it up.

The part of my job that had me concerned—and by "concerned" I mean "petrified"—was replacing the solar array. Hubble's solar arrays use photovoltaic cells to collect energy from the sun. That energy is channeled through diode boxes that convert it to electrical power, which is then stored in the telescope's batteries. From the batteries, the power is distributed through the PCU, which acts like the electrical grid that parcels out energy to the different houses and buildings on a city block.

The larger a solar array is, the more energy you can capture with it. The challenge NASA faced when Hubble launched was that photovoltaic technology wasn't that advanced. They needed arrays with a large surface area, but those arrays needed to be small and lightweight for launch. The array that was used was made out of thin, flimsy metal that stored compactly and rolled out like a window shade. The problem with that design is that, as the telescope orbits from day to night and from night back to day, there's a 400-degree swing in the temperature. It's called the thermal cycle, day-night, hot-cold. These flimsy arrays started expanding and contracting with the thermal cycle. They were shaking the telescope and getting bent out of shape and becoming less efficient.

By the time we launched Servicing Mission 3B, photovoltaic technology had improved dramatically, and the arrays we were bringing up were much smaller than the originals: one-third the size while producing 20 percent more power. They weren't flimsy; they were rigid and made out of a strong aluminum-lithium alloy. They didn't roll out like a window shade; they opened like a book, with two panels that hinged around a central mast—the spine of the book. The mast held the connectors that attached the array to the diode box in the telescope.

For launch, the arrays would be stored in their folded position inside a carrier in the payload bay. One of my jobs would be to remove the array from its carrier, after which I had to rotate it 180 degrees along its long axis in order for the mast to be in the right position to get plugged in to the telescope. I wouldn't be lifting the array out of its carrier—I would be holding it while Nancy, using the robot arm from inside the shuttle, lifted me and it together out of the payload bay. Once I was clear of anything I might bump into, I would rotate the array to put the mast in the right position. Linnehan had to perform the same task with the starboard solar array on EVA #1.

Here was my problem: This solar array weighed 640 pounds, and even though it wouldn't have any weight in space, it would still have mass, which means it still had inertia. And because it had this bulky mast on one end, the center of the array's mass was not in the center of the array; I couldn't pivot it around the middle. And even though this array was smaller than the old one, it was still enormous. When folded in half, the way it would be when I rotated it, it was 8 feet by 12.375 feet—about one and a half times bigger than a king-size mattress. The center of mass would be far away from me, making the array difficult to control. On top of *that*, I couldn't be tethered to it. This thing was big enough and had enough mass that NASA was worried if it got away from me, it might break my safety tether and take me with it or rip a hole in my suit, and they would rather lose a solar array than lose an astronaut.

No astronaut had ever done this before. The first person to attempt it would be Rick Linnehan, who had to perform the same task the day before me. I don't know if Rick was as anxious as I was, but I was terrified I was going to lose control of this thing. If I gave it the slightest jerk or moved it too fast and it started to wobble and get away from me, it would be "Bye-bye, solar array." And it's not like you can say, "Oh, let me pop down to the payload bay and get the spare." There're no spares. I would have one chance to do it perfectly.

In space, there are no small mistakes. Every mistake is a big mistake, and I'd seen astronauts make them. One robot arm operator, while trying to grapple a satellite, accidentally tipped it instead, sending it spinning out of orbit. The shuttle commander had to jump into action and maneuver the shuttle to chase the thing down. Another time a spacewalker—and this is a true story—accidentally put his right boot on his left foot and put his left boot on his right foot. Once he got outside, he couldn't fit in a foot restraint. Another spacewalker accidentally went out with a used CO_2 scrubber in his suit; he got a "high CO_2" alert in a matter of minutes, and the whole EVA had to be called off.

Mistakes cost time, and time is very expensive in space. The total budgeted cost of STS-109's eleven-day mission was $172 million—about $650,000 an hour. The solar array that I was petrified I was going to send sailing off into space? It was worth nearly $10 million. The Advanced Camera for Surveys that Newman and I had to install? That instrument alone cost $76 million. And the Hubble itself is priceless. The decades of work that have gone into it, what it does for science and the advancement of human knowledge, you cannot begin to put a value on that. And NASA was entrusting it to me, the guy who'd never been to space.

There's an old NASA saying that Newman taught me: "No matter how bad things appear," he said, "remember, *you* can always make them worse." It's true. Once a problem comes up, if you panic or act too fast, you will only exacerbate the problem. The same way

I was scared I was going to FOD the jet when I first flew in the T-38, I was in a constant state of worry that I'd be the guy making things worse.

Fortunately, if there's one thing NASA knows how to do, it's condition people to deal with fear. No training experience on Earth can ever re-create exactly how it feels to be in space. So what NASA does is, they break the experience of spaceflight and spacewalking down to their constituent parts. You work on each one individually and then piece them together. That's the way it is for all the elements of the flight, whether it's working the robot arm or working the shuttle systems or learning how to use the toilet.

For spacewalking, we have the pool. That's the major training tool because that's where the experience is as close as it will be in orbit. The shuttle's payload bay is sixty feet long and fifteen feet in diameter. In the pool there's a replica of it in the exact same configuration you'll have for the mission. For STS-109, the Hubble was going to be berthed on a rotating turntable, like a lazy Susan, at the far end of the payload bay. In between it and the airlock were the enclosures that housed our tools and equipment as well as the carriers that held the instruments we were about to install. That payload bay mock-up in the pool is a good simulation of the working environment we have in orbit, but many elements simply aren't the same. Water creates drag. If you lose control of an object in the pool, it will eventually slow down and stop; if you lose control of an object in space, it will keep going and going and going and going.

Another thing that's different is the visual. The mock-up is not the same as the actual telescope, because it's made for the pool. The actual equipment is so sensitive it can't be put in the water, so it has to be a bit different. To work with the real equipment, we'd go to the Goddard Space Flight Center in Maryland. There they have what's called a clean room, a room with a positive airflow so that no dust can ever form; it's tested down to one part per billion. Just to get in I'd have to take an air shower to blow off the dirt and loose skin on my clothes. Then I'd have to put on a gown, a hood, a mask, gloves, and booties over my shoes. Then I'd walk through this airlock into a gigantic, warehouse-size room with guys in bunny suits walking around with clipboards and cranes moving equipment overhead. It felt like a scene out of a James Bond movie.

In the clean room they have a high-fidelity, life-size mock-up of the telescope, a perfect replica: the exact same instruments, how they feel, what they look like. It's especially accurate on the inside, down to the intricate switches and the latches and the connector pins. The tools we used there were the exact same as the tools we would be using in space. In the clean room, we'd work with this replica, mating the new solar array to the telescope, aligning and installing the ACS. We'd memorize what everything looked like, how the pins and connectors lined up, how they fit together. We'd practice over and over and over again until we could do it blind-folded.

The downside at Goddard is that we were in regular gravity. The EVA suit weighs over 200 pounds. The solar array weighs 640 pounds. We couldn't actually move any of this equipment around the way we would need to in space. To practice mass handling, we went to a virtual-reality lab. There, we had a machine we called Charlotte because it looked like an enormous spider in a web. It was a box with different handrails and wires coming off it. I'd put on the virtual-reality helmet and move the handrails around; they were programmed to behave as if I were manipulating a 640-pound king-size mattress in the vacuum of space, where the tiniest misstep could send the thing wobbling out of control.

Of course, handling something in virtual reality isn't exactly the same as handling an actual physical object. For that we had what's called the air-bearing floor, which works like an air-hockey table in reverse. It's a floor that's polished to be perfectly flat and smooth. Instead of the floor shooting air up, it has objects that glide on the surface like a magic carpet by pushing air down, creating a frictionless, weightless environment. I could take an object like the solar array, put it on the air-bearing floor, and move it around in two dimensions, X and Y, and feel how easily I could lose control of it.

Not one of these training exercises comes close to the real thing. Each one mimics a certain aspect of being in space. I'd work with the real equipment in the clean room at Goddard and get a sense of what it was going to look like. Then I'd file that memory away. I'd play out the scenario in virtual reality and get a sense of how the mass handling was going to feel. Then I'd file that memory away. I'd do it again on the air-bearing floor and file that memory away. Then, piece by piece, I was synthesizing that information together into a mental model of what the experience was going to be like once I was in space.

So that's what I did. I would get in the virtual-reality lab first thing in the morning and slowly, *slowly*, rotate that array: Right hand moves an inch. Left hand moves an inch. Right hand moves an inch. Left hand moves an inch. Then I'd rotate it on the air-bearing floor: Right hand moves an inch. Left hand moves an inch. Right hand moves an inch. Left hand moves an inch. I'd rotate it in the pool: Right hand moves an inch. Left hand moves an inch. Right hand moves an inch. Left hand moves an inch. For months and months and months. Over and over and over.

. . .

Grunsfeld used to say that the Hubble knows when it's about to get fixed, and it breaks something else so you can come fix that, too. Sure enough, on November 10, three months before we were set to launch, one of the telescope's reaction wheels conked out. It was decided that Newman and I would handle swapping out the old reaction wheel for a new one after replacing the solar array. Our launch date was pushed back one week, from February 21 to February 28, to give us time to train on the new task.

With each passing day the size and the scope of the mission grew bigger. The two EVA teams trained together for months. Then Nancy joined us on the robot arm to practice flying us around in the pool. Then Scooter and Digger joined in, and we had the whole crew do stand-alone sims for ascent, entry, and orbit. We practiced different aborts and failures and contingencies. Over and over and over again.

In the early going, during our sims, we usually had an instructor in a chair with a binder, acting like Mission Control. Then, a few months out, we were assigned our flight director and flight control teams; there's an ascent and entry team and three orbit teams that handle the three eight-hour shifts of the twenty-four-hour day. We selected our family escorts. We were assigned CAPCOMs, the person in the control room who speaks directly to the crew, the liaison between the astronauts and Mission Control. With our extended team in place, we started running integrated sims. Finally, the entire Johnson Space Center is working together: The flight control team is in the Mission Control Center, the flight deck crew is in the shuttle simulator, the EVA team is in the pool at the NBL a couple of miles down the road, and everyone is linked up via radio to execute the sim. Everyone comes together and forms this cohesive, coordinated unit.

On December 17, 2001, *Endeavor* and the crew of STS-108 landed safely at Kennedy after their International Space Station assembly and supply mission. As soon as they touched down, we were up next. We were designated prime crew. Prime crew gets everything. We were first in line for T-38 flights. We had our own prime crew quarters at Kennedy with our names on the door. From that point until launch, everything revolved around us.

At the end of January, we flew down to Kennedy for TCDT, the Terminal Countdown Demonstration Test, the final stage of preparation. The TCDT is huge. We flew down in our T-38s. *Columbia* was out at the launchpad, all decked out the way we were going to fly it, the full stack, with the external tank and the solid rockets. Anytime you fly down with your shuttle on the launchpad, you request a flyby. You get in low and do a 360 around your space shuttle. You bank around, look at it from above, and then land. It's really, really cool. We went through evacuation drills to prepare for an abort on the launchpad. We suited up and went through a full launch simulation. Except for the fact that there was no fuel in the tank, everything was exactly how it would be for launch.

The press was there, and we spoke to them for the first time. Suddenly everything felt real. The instruments and tools and carriers we'd been working with at Goddard had now been moved to the Kennedy Space Center, ready to get packed into the payload bay. When we went out and inspected the shuttle, I looked at it and I realized: *This is my spaceship. There have been spaceships on this launchpad going back to Mercury and Apollo, but this one is mine.*

As we got closer to launch I could feel this huge apparatus, this giant NASA machine, coming to life around me. Thousands of people were working nonstop, around the clock, from Houston to Florida to Maryland and a dozen other places. And all that time and energy and effort was dedicated to one task: putting STS-109 in orbit. All eyes were focused on me and my six crewmates. During my space walks, those people would be focused on me and one other person. During some tasks, like rotating the solar array, all eyes at NASA, and the attention of every astronomer and hard-core space enthusiast in the world, would be focused on me and only me. It was both humbling and terrifying.

There were times I felt completely overwhelmed. Going to space had been my dream for so long, sometimes I felt like it might *still* be a dream, like I was going to wake up and realize I was just an average Joe going to work back on Long Island in my tie and my white shirts. That was maybe the hardest thing for me: accepting that this was real. Physically I was in the best shape of my life. Getting my mind right was far more difficult.

It's called imposter syndrome, the fear that people are going to figure out that you don't belong, that you don't know what you're doing. You're afraid that one day somebody's going to tap you on the shoulder and say, "Mike Massimino? Yeah, there's been a mistake. We meant to pick the other guy." It's natural to have those thoughts, but too often I'd let them get in my way. Because I was the rookie and the youngest person on the crew, I fell into the role of being everybody's kid brother, asking questions and letting others take the lead and show me what to do. I wanted to be humble, never arrogant, which is a trait astronauts despise. But the downside of that is that I'd slipped into a subordinate role. When it was time to step up and be a leader, I wasn't prepared.

Even being the junior spacewalker, I still had to be in charge of my own tasks. I had to be confident and comfortable making decisions on the fly, telling my crewmates what I needed them to do when I needed them to do it. That kind of leadership didn't come easily for me. Sometimes during sims I would get caught up in something and I'd be so worried about making a mistake or a bad decision that I'd end up making a mistake or a bad decision. I was so concerned about being a rookie and accidentally breaking something that I tried to make up for it by studying and asking questions constantly. At times I went overboard in that regard. Asking too many questions betrayed my lack of confidence and gave some people the impression that I wasn't prepared and didn't know what I was doing. I might have sabotaged myself completely but for the great friends and mentors around me. Grunsfeld told me as much during one of my evaluations. He said, "Mass, I believe in you, and I believe that you can do this. Your problem is that you don't believe in yourself."

One evening right before launch, Steve Smith came over to my house to talk. I'm sure he could tell what I was feeling. He said, "Mass, I want you to remember two things. One: Know that you're prepared. You may not feel like you're prepared, but they wouldn't let you go if you weren't. And two: Space is an open-book exam. You're not alone up there. This is a team, and you can always get help if you need it."

That night was a real turning point for me. John and Steve were both right. I had to stop thinking of myself as a rookie. I was not a rookie in the eyes of Hubble engineers and astronomers and management and instructors—I was one of the guys who was going to fix the Hubble. I couldn't leave my responsibilities to my crewmates. I was fully capable and, more important, during eighteen months of training, I had demonstrated that I was fully capable. Everyone from the NASA administrator to our janitorial support in the NBL had confidence in me; now, for the sake the team, I needed to have the same confidence in myself. Accepting that and knowing that was probably the hardest part of preparing for the mission.

. . .

Once you've got your body and your mind right to go to space, there's only one thing left to deal with, something you don't hear engineers and scientists talk about a whole lot. You have to prepare your soul. Being ready to go means being ready to *go*. You're not just preparing to leave the Earth for two weeks—you're preparing for the possibility that you might be leaving forever, and you have to be at peace with that.

Most days, you don't think about being killed. You go to work, go shopping, go home, and it doesn't cross your mind. But spaceflight is still a dangerous business. As we were counting down to launch, I thought about death constantly. I found myself having random moments. I'd linger in rooms more than normal, looking around, wondering if this was the last time I'd see those people. I took care of the will and the life insurance. I made sure the cars were washed. I made sure Carola knew where the spare key to the garage was.

Every day I was reminded of my mortality in different ways, big and small. People love having stuff that's been flown in space, and every astronaut has a kit in which we're allowed to take items up to give away. I took a photo of the students from my kids' elementary school, an FDNY patch and hat to honor my father. The Mets gave me a jersey. I was going around to my relatives and in-laws, too, asking if people wanted me to take anything for them. I was expecting them to offer a watch or some cuff links or a family photo or something. Not my family. They're Italian-American Roman Catholics. Every single one of them came in with some religious object. I'm going to space and I've got a statuette of the Madonna and child. I've got a baby Jesus from a nativity set of some cousin over in Sicily. I've got a St. Christopher medal, a St. Michael medal,

a picture of Padre Pio, one of Our Lady of Loreto. I've got cruci-
fixes, rosaries, prayer cards, all these trinkets. It got to be funny,
but there was a reason behind it: Everyone was worried I was going
to die, and if it turned out this was my time, they wanted to make
sure I was covered.

I've always been a decent Catholic. Not the best, not the worst,
but middle-of-the-road. In those last weeks I became the most de-
vout Catholic in the state of Texas. I probably went a little overboard with it. As the launch got closer, I started going to confession
a couple times a week. Right before quarantine I went to our priest,
Father Dominic. "Father," I said, "I've been coming here because I
need a clean soul, but now I have to go into quarantine and I won't
see you anymore. What happens if something happens after I go in
but before I go up? Can I send you confession via e-mail?"

I'm sure he thought I was losing my mind. He was like, "Sure,
Mike. If you want to send me an e-mail, that's fine."

The whole crew is grappling with the same fears. Everybody
deals with it in different ways. Some people throw themselves into
the work. Some people go to the gym. Different people need different things. For me, I needed to spend time with my family. That's
what was important to me as the clock ticked down in those final
weeks. I made sure I was home for dinner. I took Gabby camping
one weekend. Their school had a Skate Night fund-raiser where we
went roller-skating. I could have spent that time going over EVA
checklists for the thousandth time, but I decided to have confidence in my training and trust that I was ready so I could have
that time for myself. But as the launch grew closer, family time was
harder and harder to come by.

To rendezvous with the Hubble, we had to catch it when it
passed directly overhead; if you get to space and your target is on
the other side of the planet, good luck catching it. For STS-109,

that meant we had to launch in the middle of the night; we'd be going to bed at 12:30 in the afternoon Houston time and waking up to start our day at 8:30 at night. In order to acclimate our bodies, a few weeks out from launch, we started a sleep shift, pushing bedtime back a bit each night and waking up a bit later each morning. The further the sleep shift moved us off a normal nine-to-five schedule, the harder and harder it was to spend time at home. I was sleeping through getting the kids ready for school in the morning. Crew activities were running into dinnertime at night. It was frustrating for me.

One week before launch, the crew goes into quarantine. Nobody wants to be dealing with a head cold or a virus in space. Adults can visit you in quarantine once they've been screened by the flight surgeon, which means you can still see your spouse and your fellow astronauts. But children under the age of eighteen can't come in. That's it. Kiss 'em and hug 'em and say good-bye. We were scheduled to start quarantine at 9:00 p.m. on February 21, a Thursday night, the same night as the annual Blue and Gold Banquet for Daniel's Cub Scout pack. It was a big deal, the last thing I would get to do with my kids, and I didn't want to miss it. It started at seven, and I could leave early to make the cutoff. I told the flight surgeon, Smith Johnston, that I was going. He said, "You've gotta be kidding me. Two hours before quarantine, in the middle of cold and flu season, and you want to walk into a room with a hundred six- and seven-year-olds blowing snot and germs everywhere? Are you out of your mind?"

I said, "I'm going."

He was skeptical, but I think he understood why it was so important. "Just don't get sick," he said.

Around 8:30 we got up to leave. I wanted to slip out quietly, but the scoutmaster got up and made an announcement: "Mike

Massimino is leaving because he's going to fly in space next week. Mike's going to Hubble. Let's wish him the best of luck!" The Cub Scouts and their families gave me a cheer. That was my big sendoff. We drove home, and I dropped Carola and the kids off so she could put them to bed. I gave my wife a hug and a kiss. I grabbed Gabby and Daniel and held them as close as I could and said good-bye.

I checked into quarantine at 9:00 sharp, but while I was unpacking and getting settled in, I realized I'd forgotten my watch. It wasn't that big a deal. I knew Carola could bring it to me the next day. But I realized maybe I could use that to ask to go back and get more time. I only lived five minutes away. I went to Scooter and said, "I need to run home real quick. I forgot something."

He said, "Mike, it's 9:30. We're officially in quarantine. Nobody's supposed to leave." I leaned on him pretty hard. He knew why I was really asking. He said, "Go. Get what you need. Come right back."

I raced home and ran inside. Carola heard me coming back in after we'd just had this big good-bye. She gave me a confused look. I said, "I just need to get something." I don't remember if I even got the watch. I went down the hall and I slipped into Gabby's room. She was in her bed in her nightgown with flowers on it. I sat down in this little chair next to her bed and I watched her sleep. I stayed there for as long as I could. Ten, fifteen minutes maybe. Then I went over to Daniel's room. He was wearing his baseball pajamas. I sat and watched him, too. I was trying to memorize their faces. I couldn't make myself go down and get back in the car. Every part of me was saying, *Don't leave. Don't go.* I knew I'd see Carola again the next day, but was this the last time I'd ever see my kids? I'd been given this amazing gift, going to space, my childhood dream come true. But what if that dream cost me everything else?

I stayed for as long as I could. Then I knew it was time to face facts. I left and drove back to quarantine. Once I got there, I was okay again. I looked around at my crewmates and I remembered: *I have a second family here. They're counting on me, too, and it's time to go to work.*

15

WEIGHTLESS

The first thing I did in orbit was my Tom Hanks routine from *Apollo 13*, taking off my helmet and floating it in front of me. Then I took my gloves off one at a time and floated those in front of me as well. I looked away to do something and when I looked back up one of the gloves was gone; it had floated off. That was rookie space lesson number one: Hold on to things. They get away from you.

I started unbuckling myself from my seat. Linnehan was already out and heading up to the window to take a picture of the external tank before it dropped away from us and burned up on reentry; this was to see if there was any external damage or loss of insulation foam that needed to be documented. I was right behind him. I had to get up to the window and take a look outside. John Glenn's view of Earth in *The Right Stuff* was the thing that had rekindled my space dream, and now, twenty years later, it was my turn to see it for myself.

We were over the Indian Ocean, which was a beautiful shade of blue with puffy white clouds sprinkled across it. I felt like I was in one of those dreams where you're magically floating above everyone else. I could see the ripples in the ocean, the horizon with the blue atmosphere in a thin, hazy line. It was like all the pictures I'd seen, only a thousand times better. I lingered for a moment, staring out. Then it was time to go to work.

The shuttle's crew compartment is small, only 2,325 cubic feet for seven people to live and work in for nearly two weeks. Up on the flight deck is where you get the amazing views, with six forward-facing windows for the pilots to fly the ship, plus two windows in the roof and two in the aft bulkhead looking out at the payload bay. In the floor of the flight deck are two hatches leading down to the somewhat claustrophobic, very utilitarian mid-deck, virtually every inch of its walls taken up by storage lockers and the gear needed to live and eat and sleep in space. On the aft wall of the mid-deck is the airlock going out to the payload bay. The airlock is a cylinder with a round, forty-inch hatch leading to a space that's about five feet in diameter and seven feet long, just enough room for two astronauts in EVA suits to wait to go out on a space walk.

Scooter and Digger were up above us on the flight deck, checking out the systems, doing engine burns to put us on the right trajectory to rendezvous with Hubble. The rest of us were busy on the mid-deck going about the tasks necessary to convert the shuttle from a rocket ship to a spaceship: setting up the toilet, the galley, the exercise bike. That takes a couple of hours, in large part because adjusting to being weightless takes so long.

From the minute I started moving around, I felt like a bull in a china shop. In the space station, astronauts can barrel themselves down the tube and get up some speed and fly like Superman. You can't do that on the shuttle. You can spin around, leap from the floor to the ceiling, but that's about it. On the first day, even doing that is difficult. Your sense of motion is all messed up. You feel crazy out of control at first, or at least I did. I'm naturally clumsy, plus I'm big, and I didn't know my own strength. I was banging into everything, knocking into people. One time I reached for something on an overhead panel and my finger accidentally banged into a wall and flipped a switch. There's switches and instruments all over the shuttle—over two thousand different displays and controls on the flight deck alone—and you don't want to go around randomly turning things on and off. That's bad. So you move slowly, awkwardly, trying to develop some sense of control. The whole process is like learning how to walk again. It's the same with your hands and fingers and fine motor skills. You go to grab something and, instead of grabbing it, you bat it away and you have to go chasing it. You're like an infant concentrating on picking up a Cheerio for the first time.

And you feel horrible, absolutely terrible. Adjusting your body to space is painful. The first thing that happens is the fluid shift. There's tons of fluid in your body: blood, plasma, water, mucus. On Earth, gravity keeps it pushed down. In space, it's free to float up to your head. Everybody's face was red and flushed and puffy. We were floating around, looking like puppets in a Mardi Gras parade with giant papier-mâché heads. The other thing that happens is that your spine elongates—again, because there's no gravity keeping it compressed. You grow about an inch in space, and all those sensitive muscles in your back have to stretch and adjust. That's painful, too.

Then there's the nausea. "Stomach awareness" is the official term. That whole first day I floated around feeling like I was going to barf at any moment. Space sickness is actually the opposite of seasickness. The effect is the same, the nausea and the vomiting, but the root cause is different. When you're below deck on a boat, you can't see the motion of the sea, so your eyes are telling your brain that you're completely still, but your vestibular system is going up and down with the waves. It's the same thing if you're trying to read in a moving car. The conflict between those two sensory inputs is what creates the feeling of nausea. In space, you're floating around and this time it's your eyes that are telling your brain that you're moving and your inner ear that's telling your brain that you're still, because your inner ear doesn't move when you're weightless.

The more you move around, the worse it gets. You think you're going to get to space and be weightless and have fun doing flips and floating upside down, but in space there is no up or down. To your brain, floating sideways or upside down feels the same as standing right side up. So if you spin around or flip upside down, the sensation you get is not that you've spun or flipped around. What you feel is that the room is spinning and flipping around you while you're staying perfectly still, which causes the worst vomit-inducing feeling of vertigo you've ever experienced. After a couple of days you get used to it. You can have a conversation on the ceiling and not notice. But it takes the brain time to adapt, so at the beginning you move as slowly as possible.

One of my main jobs that first day was to help Nancy set up the robot arm and open the payload doors, an important task. The equipment on the shuttle generates heat, and the payload bay doors have radiators on them that radiate that heat out into space; otherwise you'd cook yourself. If you can't open those doors, you're going home. I tried to ignore the nausea and focus on doing that with her. Everyone else was doing the same, plodding around, doing their tasks, nobody saying much. It was not a party atmosphere. It was not "Yeah! We made it to space!" It was "Ugh. Leave me alone. I'm gonna puke." That kind of sickness trumps everything. You can be in space, you can be at Disney World, but as long as you have that grumpy, barfy, nauseated feeling, nothing is going to make life okay. I forced myself to drink a bunch of water and immediately threw it up. After that, I felt better.

Around six hours post-launch the shuttle was set up and ready for our journey to Hubble. I had a bit of time to look out the windows, but not much. It was already time to go to bed. Window shades go up on the flight deck, because daybreak comes every ninety-seven minutes and you have to block out the sun. Once they were installed, we started winding down. I took off my contacts and put on my glasses. I brushed my teeth. When you rinse, you either have to spit into a towel or swallow; I swallowed. I took a sleeping pill to help me get to bed.

Your first night in space is weird. The pilot and the commander sleep on the flight deck. Other than that, you can sleep pretty much anywhere. On the ceiling if you like. Grunsfeld wanted to sleep in the airlock, because it's cooler in there and he likes being cold when he sleeps. I was in the mid-deck with everyone else. You have a sleeping bag and these clips you use to attach and cinch it to the wall. You don't want to be floating around because you'll knock your head. Then, when you get inside the sleeping bag, you're kinda floating inside this cocoon. Once you get used to it, it's the most relaxing way to sleep ever. What they've also found over the years is that people like having their head against something, even if they're floating in the air, so NASA developed pillows that attach to your head with a Velcro headband.

I felt strange that first night. I've never dealt well with transition, with new things, and this was the ultimate new thing. I was out of sorts. I felt horrible. Everyone was grumpy. I was like, *Is this all there is?* But then once I was in bed—and maybe it was the anti-nausea medication—after a few minutes I started to feel better. There was something about going to bed that made me feel okay. The day was over and I had a chance to relax and reflect. I'd made it to space.

With my copilot, Snoopy, July 1969.
Backyard adventures in space.

With Mom and Dad at my Columbia
graduation. They are smiling because
they thought my formal education was
over; little did they know . . .

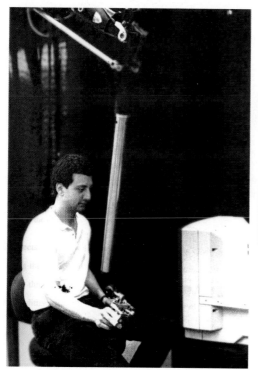

Controlling the robot arm in my lab
at MIT, with a 1980s hairstyle.

I'm on the left with (left to right)
Swiss astronaut Claude Nicollier and
colleagues Mike Meschler and
Lonnie Cundieff during a test of our
manipulator position display at the
Johnson Space Center in 1994.

With one-year-old Gabby on my shoulders
and space shuttle *Columbia* on top of a 747 at
Ellington Field, 1994.

With eight-month-old Daniel
at Jekyll Island, Georgia,
Spring Break 1996.

A very happy pose for my first official
astronaut photograph. BELOW: Flying
at what we called "the speed of heat"
in a NASA T-38 above the
Gulf of Mexico.

All dressed up and about to be lowered
into the water for a challenging day
of spacewalk training at the Neutral
Buoyancy Laboratory (NBL).

The NASA astronaut class of 1996. The Sardines—thirty-five Americans and nine international astronauts, the largest (and best-looking) astronaut class ever. (Can you find me?)

With Dad, Mom, Carola, Gabby, and Daniel, at the Sardine astronaut class graduation, from astronaut candidates to official astronauts, April 1998.

With Mike "Bueno" Good (on robot arm) training in NBL. Divers on the right are holding an IMAX camera and filming the documentary *IMAX: Hubble 3D*.

Checking out the payload bay of space shuttle *Columbia* with Rick Linnehan (behind me). Our space-walking instructor Dana Weigel is next to Rick and apparently falling asleep after a late night at the Cape.

Team photo of Cold Lakes, Canada. Back row of astronauts: Greg Chamitoff, Lee Morin, Frank Caldeiro, Charlie Precourt, Dan Tani, and me. Front row of Canadian Army instructor tough guys: Sgt. Colin Norris (with impressive facial hair) is on the left.

The STS-109 taking a break from emergency training at the Kennedy Space Center. Space shuttle *Columbia* on the launchpad is in the background. Top row (left to right): John Grunsfeld, Scott Altman, Nancy Currie, and Jim Newman. Bottom row (left to right): Duane Carey, Rick Linnehan, and me.

The Hubble Space Telescope on the shuttle's robot arm as seen through an overhead window in the space shuttle. Beautiful planet Earth in the background.

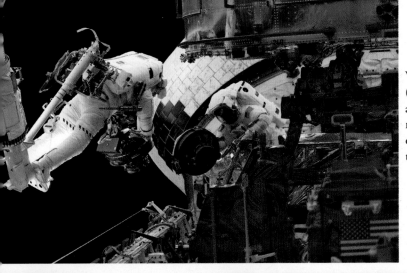

With Jim Newman (on left) to exchange an old for a new reaction wheel during our first spacewalk together.

Trying to look relaxed and cool for my spacewalking hero photo taken just minutes into my first spacewalk on STS-109. Notice the Earth in my visor.

Happy moment with Jim Newman after successfully completing our first spacewalk.

Rick Linnehan on the arm rotating a solar array and trying to not be distracted by the view of the Earth in front of him. My turn came the next day during nightfall, so there was no worry about such distraction.

Another boyhood dream fulfilled: ceremonial first pitch from the mound at Shea Stadium before the Mets versus Yankees subway series game, June 15, 2002.

The crew of STS-107 on orbit just days before they did not make it back to Earth onboard space shuttle *Columbia*.

Two space shuttles ready to go on the launchpad for the first time in history. Our spaceship, *Atlantis,* is in the foreground, and our rescue spaceship, *Endeavour,* is in the background along with a lucky rainbow in the sky.

The crew of STS-125 outside our ride to the launchpad on launch day
(left to right): me, Mike Good, Drew Feustel, John Grunsfeld,
Megan McArthur, Greg Johnson, and Scott Altman.

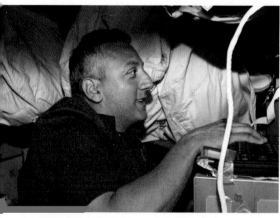

At the computer sending one
of the first tweets from space to
planet Earth.

With Drew Feustel in the airlock
of *Atlantis,* wishing each other a
last-minute good luck.

Space shuttle *Atlantis* lifts off on its mission to unlock the secrets of the
universe, May 11, 2009, at 2:01 p.m. Eastern Daylight Time.

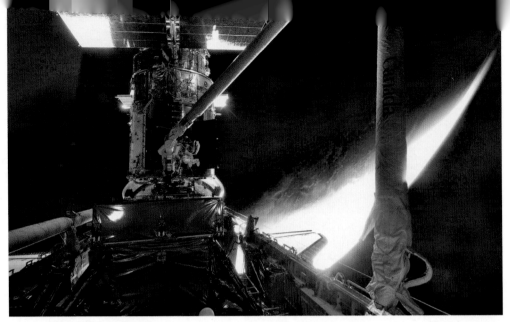

Going into a sunrise during the fourth and final spacewalk of my astronaut career.

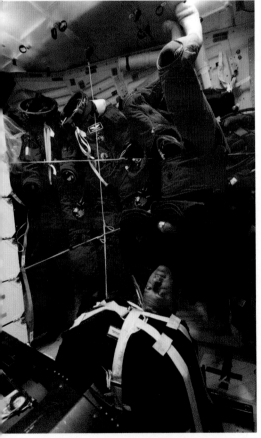

With my copilot, Snoopy, May 2009.
Same Snoopy, but now with real
adventures in space.

Having mixed feelings as I prepare for
entry and landing on my last day in
space, May 24, 2009.

Every couple of hours that night I woke up. The first time I didn't know where I was. That feeling you get when you wake up in a strange hotel room and for a minute you don't remember what you're doing there—it was that same feeling, only in space. *Where am I? Where are my kids? This isn't my house.* The weird thing about waking up in the middle of the night on the shuttle is that it's the only time you're actually alone in space. The rest of the day you've got six other people on top of you. But you wake up and everyone's asleep in their cocoons. You're groggy and fuzzy and there's nothing but the dim light from the toilet, the low hum from the fans, every now and then a crackle from the radio upstairs. You could easily be stranded alone on a spaceship orbiting Mars in a science fiction movie. It's very, very eerie but also really, really cool.

Every morning on the shuttle the ground crew plays wake-up music selected for members of the crew. We woke up on day two to John Hiatt's "Blue Telescope." By that morning I was hungry again. I ate a peanut butter and jelly sandwich I had left over from the day before, and once I started eating, for the next ten days I didn't stop. Eating in space was fun. All your food is pre-prepared. You don't have to cook it. It's dehydrated and you add water and heat it up. You select your own menu, too. Spaghetti and meatballs, macaroni and cheese, shrimp cocktail, steak, lasagna. The hot meals are in pouches, and you cut the pouch open and eat. You have to be careful, because everything floats, but that's the fun part. Popping M&M's in the air and going after them and chomping them like Pac-Man. I actually gained weight in space, which no one ever does. The doctors were confounded, but I just loved eating up there.

The drinking water on the shuttle is a by-product of the fuel cells. It's actually a brilliant piece of engineering. We have tanks of liquid hydrogen and tanks of liquid oxygen. When they're combined it creates a reaction that produces power for the fuel cells, and the by-product of that is water, which is then purified with iodine. It's much better than the drinking water system on the space station. They use solar arrays for power, which means there's no water being generated. They have some delivered, but 80 percent of it is recycled urine, sweat, and condensation collected through a filtration system that cleans it and puts it back out. As my pal Don Pettit described it: Today's coffee is tomorrow's coffee. I was glad I was going to Hubble.

By day three, we were closing in on Hubble and our main objective was to rendezvous with it, a complicated and delicate task. Naturally, we were roused by the theme from *Mission: Impossible*. Wake-up was around 8:30 p.m. Houston time. Just after midnight we reached the telescope's orbit, coming in about ten miles behind it. We slowly closed the distance between us until finally we had visual contact. At first sight, the sun reflecting off the telescope looked like a distant star, another point of light among all the others. Slowly it grew bigger and bigger, taking the familiar shape I was used to: a bright silver cylinder. It was shinier than I expected. This man-made thing, this marvel of human engineering out in the middle of space, was an incredible thing to witness.

The final half mile of the approach was hand-flown, with Scooter taking manual control of the shuttle, firing the engines to slow us down and make tiny course corrections. Nancy was preparing the robot arm. I was standing by as her backup and also taking photographs and pictures to document the rendezvous. The whole crew was tense, focused. Scooter was closing the distance between us at less than half a mile an hour. The whole thing played out like the high-tension climax of an action movie, only in slow motion. At thirty-five feet, Scooter held our position and handed things over to Nancy to grapple the telescope and bring it in.

At 3:31 a.m., over the Pacific Ocean just south of the coast of

Mexico, Nancy successfully grappled the Hubble and brought it down and secured it in the payload bay outside the cabin. There was a huge sigh of relief and much rejoicing from everyone in the cabin. We were happy campers. But relief gave way to anxiety as a new thought overwhelmed me: Now that we had it, I was going to have to go out there and spacewalk on it. My mind started racing with thoughts and worries about what I was going to have to do. To be honest, once I got used to being in the shuttle, even though I was in space, it didn't seem like that big a deal. Other than the weightlessness and the view, you could easily be inside the simulator back in Houston. It's actually quite comfortable. The food is good. The toilet works. We were wearing polo shirts, for Pete's sake. How dangerous could it be? But spacewalking was a different story. Spacewalking was going to be nuts. We were going to get into these suits and put on helmets and gloves and *go outside*.

On day four we woke up to "Five Variations on Twinkle, Twinkle Little Star." This was Grunsfeld and Linnehan's big day. Whichever spacewalking team isn't going out works as backup for the team that is. You help them get dressed, check their gear, like the cornerman helping a fighter get ready for the ring. Once the other team is outside, you and your partner take turns running the EVA checklist. The spacewalkers have enough to do without having to worry about remembering each and every task.

That first space walk was going to be a mirror image of what Newman and I would be doing the next day, swapping out a solar array. I thought it would be good idea for me to watch them do it first and see what I could learn. That was a bad idea. I would have been better off just going out and doing it. I was watching those guys out there in front of me and thinking, *Oh my God. They're actually outside. Spacewalking. I'm gonna have to do that?* It scared me to death. Linnehan had the same daunting task I did, rotating the array, and it was his first space walk, too. There were a few moments where it went herky-jerky on him and he had to bring it under control. I winced every time. *I'm doomed,* I thought.

As nerve-racking as it was to watch from inside, that first space walk went smoothly. The new array went on without any problems, but the hard work took a toll. After they came back in, I went to Linnehan. He looked exhausted, physically and mentally. He was drenched with sweat, his fingers and hands white and pruney from the moisture. His hair was messed up and he had red marks all over his skin from rubbing against his suit He looked like he had been through a war and come out the other side. I asked him, "What was it like out there?"

"It was hard," he said. "Much harder than in the pool."

Some people will tell you the pool is actually harder, because there's resistance and gravity and other forces to contend with. "It'll be easier in space," they say. Oh no it won't. Why? Because you're in space, that's why. Everything is harder in space.

That night Newman and I prepped for our space walk. One of the things you do is put anti-fog on your visor. It's Joy soap, actually, the kind you buy at the supermarket; it just happens to work well as an anti-fog solution in space suits. Joy stopped making this particular kind, so NASA bought up a lifetime supply, basically every bottle available in the world. You have an applicator and you rub this soap on and buff it. Newman and I went up on the flight deck and watched the Earth go by in the windows while we polished and buffed our helmets.

I took another sleeping pill to get to sleep that night. I was that nervous. I knew that the next day was my day. It was going to happen. I was going to go out in that space suit and I was going to have to perform. I'd be doing it under a microscope, too. Everything you do out there is being recorded by helmet camera. Everybody's watching. Anything goes wrong, and everybody knows. I had this incredible anticipation. Not like Christmas morning; more like the first day of school, where you're excited by the new possibilities but also terrified about making friends and not screwing up.

The next morning I woke up, had breakfast. I put on my polypropylene underwear to absorb my sweat, my liquid cooling garment, the biomedical sensors the ground crew would be using to monitor my every breath. I got my drink bag ready; you can't have any air bubbles in it, so you spin it around until the bubbles are at the top and then you squeeze those out. Grunsfeld and Linnehan helped Newman and me into our suits, first the pants, then the torso, then the gloves. I went over my notes in my flight notebook, went over my checklist. The final step was the helmet. I scratched my nose one last time, gave a nod, and Grunsfeld carefully placed the helmet over my head, lowered it onto the neck ring, snapped it into place, and then locked it.

Inside the airlock, you go through your final checks and then you to do a forty-minute pre-breathe of pure oxygen. The air we breathe on Earth is a combination of nitrogen, oxygen, and other gases, and the air pressure at sea level is 14.7 pounds per square inch (psi). When your body moves to a lower air pressure, like the vacuum of space, you can get nitrogen bubbles forming in your blood, which is what causes decompression sickness—the bends. The atmosphere and air pressure inside the shuttle are normally engineered to be identical to what we experience on Earth. But twenty-four hours before the first space walk we depressed the shuttle cabin to 10.2 psi and kept it there. That made the change in air pressure less extreme. Then you do the pre-breathe of pure oxygen on top of that to rid your body of nitrogen, and that way you don't get sick.

During the pre-breathe, you're attached to the wall to keep you from banging around. So, right before you're about to face the most difficult moment of your entire life, you've got forty minutes to do nothing but hang there and obsess over everything that could go wrong. I tried to stay focused, going over the checklist on my cuff, thinking about my tasks, double- and triple-checking to see if everything was right with the suit. But my mind wandered. My eyes kept drifting nervously. At one point I looked at Newman and we locked eyes. He nodded at me and I nodded at him. Then I looked over at the outer hatch. I remember staring at it and thinking: *There it is. That's the door to space.*

I wonder what's on the other side.

16

EARTH IS A PLANET

When people ask me what it feels like the first time you space-walk, what I tell them is this: Imagine you've been tapped to be the starting pitcher in game seven of the World Series. Fifty thousand screaming fans in the seats, millions of people watching around the world, and you're in the bullpen waiting to go out. But you've never actually played baseball before. You've never set foot on a baseball diamond before. You've spent time in the batting cages. You've run drills and exercises with mock-ups and replicas. You've spent months playing MLB on your Sony PlayStation, but you've never once set foot on that mound. And guess what? The Series is tied and the whole season is on the line and everyone is banking everything on *you*. Now go get 'em. That's how I felt sitting in that airlock. NASA was trusting me to do millions of dollars in repairs to this billion-dollar telescope . . . and until that moment I'd never laid a hand on the actual telescope.

It was a moment I'd dreamed of my whole life, something I'd worked toward for years, yet I couldn't help but worry: What if I was no good at it? What if I didn't actually enjoy it? What if I hated it? If Mission Control had come over the radio at that moment and said, "Hey, we just realized the Hubble's fine and you don't have to go out," part of me would have been relieved.

I watched the clock tick down, anxiously waiting for it to get to zero. Finally it did. Once the pre-breathe was over, we unhooked our suits and we were floating. Scooter came by for a handshake and then Digger popped in for one last good-bye. During our training, as the only two rookies, Digger and I had become close. In the months leading up to the flight, we talked about dreaming about space since we were kids and what we were most looking forward to. Being weightless and looking out the window excited us the most, but I was going to get to spacewalk and he wasn't. Right before we launched he'd come up to me and said, "Mass, since I'm a pilot I'll never get a chance to spacewalk, but you gotta do something for me. I want you to look around out there and, as soon as you get in, I'm going to come to you and I want you to tell me what it's like. I want a description fresh from your mind. You gotta promise me you'll do that." Now he wanted to make sure I made good on my promise. "Good luck, Mass," he said. "You'll do great, and remember, I want a full report."

I told him I would give him one, and quietly I thought to myself: *I hope it's a good one.* Then Grunsfeld floated over to the airlock's inner hatch and pushed it closed. He pulled the handle down and spun it shut. It sounded like I was being locked in a prison cell. *WHOMP! CHA-CHUNK!* I looked over at Newman like *I guess this is it. There's no going back now.*

Newman switched our suits over to their own battery power and oxygen. Then he started to depress the airlock. After a final purge of air from the airlock, we were in a complete vacuum. There was no sound. The *cha-chunk* I heard when they locked us in, I wouldn't hear that now. I could bang on the shuttle wall with a hammer and I wouldn't hear that, either; there's no way for that sound to travel. The only noise I could hear was the sound of my internal cooling fan and some squeaking from moving around inside the space suit. My voice sounded different, too, because the sound wave travels differently through the lower atmospheric pressure. It's at a lower register. I sounded like I was about to cut a blues album.

At that point we were clear to go. There was nothing left except for Newman to open the outer hatch. Once he did that, I knew the only thing between me and certain death would be this suit I had on. NASA's EVA suit is amazing. It's got its own life support and it's pressurized and it's got a Kevlar lining. It's designed using the most advanced technology known to man. But still, it's just a suit. I could get a hole in it, either from getting punctured by a micro-meteorite or from the sharp edge of a tool. Or sometimes suits can leak. If the hole was large enough and I didn't get back inside before my suit pressure got to zero, I'd die. My tether might break and I'd go tumbling off into space and my crew might not get to me before my life support ran out. It's also possible to drown in a suit. If the water and cooling systems malfunction, fluid can leak into the suit, which has actually happened. Luckily the spacewalker got back inside in time. Ultimately you have to have trust in your equipment and the people who built it for you. You say to yourself, *The suit will work*, and you try not to think about the alternative. That's a key part of your training, too. You build up so much confidence in your gear and in the people who handle it for you that when it comes time to go out, hopefully you don't even think about it.

Newman pulled open the door to the payload bay and pushed the thermal cover aside. Then he went out first to make sure the coast was clear and to secure our safety tethers. He was out there for a few minutes. Finally he said, "Okay, you're clear to come out." I put my hands on the hatch frame and pulled myself through. I was floating on my back, looking up and out of the payload bay. The first thing I saw was Newman floating above me, hanging out with this grin on his face like *Check this out.*

Behind his head was Africa.

According to my suit's biometric sensors, I have a normal resting heart rate of 50 or 60 beats per minute. The moment I saw the Earth it spiked to 120. Eventually it settled back down to a normal exercise rate of around 75, but for that moment it was racing. Hubble is 350 miles above Earth; we need the telescope as far away from the planet as possible in order to have a longer orbit, see more of the sky, and be farther away from the atmospheric effects of Earth. The space station is 250 miles above Earth. From that vantage you can't fit the whole planet in your field of vision. From Hubble you can see *the whole thing.* You can see the curvature of the Earth. You can see this gigantic, bright blue marble set against the blackness of space, and it's the most magnificent and incredible thing you've ever seen in your life. One thing I was not prepared for was how blue it is, how much water there is. Rick Mastracchio, a veteran spacewalker, once described it to me by saying, "You're always over the Pacific." And it's true. The Earth is three-quarters water, and it feels like it. You're up there and it's water, water, water. Then there's Africa and, *poof,* it's gone in a few minutes and then it's water, water, water again.

Seeing the Earth framed through the shuttle's small windows versus seeing it from outside was like the difference between looking at fish in an aquarium versus scuba diving on the Great Barrier Reef. I wasn't constrained by any frame. The glass of my helmet was polished crystal clear, and every direction I looked, there was nothing around me but the infinity of the universe. I was really out there, floating in it, swimming in it. I felt like a real spaceman. When you're orbiting the Earth on an EVA, you're in your space suit flying at 17,500 miles per hour. You're moving fast, but it doesn't feel like you're moving at all. You're falling, really. That's what orbiting is: You're falling toward the Earth but you're also moving so fast parallel to the planet that the edge of the Earth keeps rotating away from you as you fall, so you keep going round and round.

Whenever you're falling, you're weightless. People often think you're weightless in space because there's zero gravity. That's a common misconception. Gravity exists everywhere, same as friction and inertia or any other property of physics. The moon has one-sixth the Earth's gravity. In space there's microgravity. But anytime you're falling, even on Earth, you are weightless. And whenever you're weightless, your vestibular system is still, which tells your brain that you're not moving. Also there's no atmosphere to create a drag on your suit, which means there's no sensation of movement whatsoever. The only way to get a sense of how fast you're going is to look at the Earth. You spend forever over the Pacific and then you hit California and you realize you're really booking. There's LA, there's Vegas, there's Phoenix, there's Albuquerque. *Bam bam bam bam.* They whoosh by. Baja to Miami is a six-hour flight. In space you do it in eleven minutes.

After seeing the Earth, I looked down the payload bay at the telescope, and the thing I noticed was the light from the sun. The sun on Earth is filtered through the atmosphere; it can appear bright yellow or as that golden hue you get at sunset. You get different colors depending on the place and the time of day, which in turn affects the color of objects as we perceive them. In space, sunlight is nothing like sunlight as you know it. It's pure whiteness. It's perfect white light. It's the whitest white you've ever seen. I felt like I had Superman vision. The colors were intense and vibrant—

the gleaming white body of the shuttle; the metallic gold of the Mylar sheets and the thermal blankets; the red, white, and blue of the American flag on my shoulder. Everything was bright and rich and beautiful. Everything had a clarity and a crispness to it. It was like I was seeing things in their purest form, like I was seeing true color for the first time.

After taking a minute to soak everything in, my first conscious thought was *How the heck am I going to get anything done? How am I supposed to pay attention to my work with this magnificent beauty all around me?* But then I turned my head back to the wall of the payload bay right in front of me and there was a handrail. I recognized it. It was just like the handrail I'd seen and worked with in the pool dozens of times. I looked around and everything in the payload bay was right where it was supposed to be. Every tool, every piece of equipment, the winch with the rope on the end of it—everything felt familiar. Even though I'd never been out on that pitcher's mound before, I knew exactly what I needed to do.

The first thing I did was pull myself up to the window looking out from the flight deck onto the payload bay to get my picture taken. The hero shot, we call it, a nice memento of your first moments of spacewalking. With that out of the way, I spent the next fifteen minutes or so doing what's known as translation adaptation. When you train for space walks, you're used to moving around in the pool, where there's drag on your suit from the resistance of the water. It slows you down and makes you more stable. In space there's no resistance to any move you make, so you have to go *really* slow. I moved up and down the forward part of the payload bay. I did some pitch and roll maneuvers, getting a sense of how it felt. I took note of how my tether was moving behind me so I could make sure I wouldn't get tangled up in it.

Then it was time to go to work. The robot-arm platform was positioned right at the front of the payload bay where we'd come out. I climbed onto the top of the robot-arm platform, slotted my boots into the foot restraints, and they clicked right into place. Flat and go. Perfect on the first try. Now I was ready for Nancy to move me around. For the next twenty minutes or so, Newman and I moved about the payload bay, getting things ready. We had a foot restraint that attached to the Hubble itself, so the free-floating spacewalker could anchor himself to work on the telescope. We hooked that up and did everything else we needed to do.

While we were doing this, we entered our first night pass. The Hubble orbits around the Earth once every ninety-seven minutes. Because it's so far out from the planet, it's exposed to the sun for most of that; approximately two-thirds of the orbit is in daylight, and one-third is nighttime. We had egressed from the airlock about midway through our first day pass, and now we were about to leave that perfect white light and plunge into complete darkness. When night comes in space, you feel it before you see it. The temperature swing from 200 degrees Fahrenheit to –200 degrees Fahrenheit occurs in an instant. The amazing thing is that your suit protects you from that; the temperature inside stays within a tolerable range, and you have a temperature control valve you can adjust to warm up or cool down as needed. So the 400-degree swing isn't harsh, but you definitely still notice it. The best I can describe it is like when you're in the ocean on a warm summer's day and a cold current rushes past and it gets you down in your bones. That's what it's like. This chill ran through me and then a few seconds later it was like somebody flipped a switch and everything went black. If the sun in space is the whitest white you've ever seen, nighttime is the blackest black. It is the complete absence of light. You have some lights in the payload bay, and you have your helmet lights that light up your work area, but you can look around you and all that white, that purity that was there, it's gone. There's nothing.

You see the stars, of course. You can see the whole universe. At night, without the sun, space becomes this magical place. In space, stars don't twinkle. Because there's no atmosphere to fog your view, they're like perfect pinpoints of light. Stars are different colors, too, not just white. They're blue, red, purple, green, yellow. And there are billions of them. The constellations look like constellations. You can make out the shapes and see what early astronomers were getting at with their descriptions. The Southern Cross was my favorite. And the moon feels like it's right there. It's not a two-dimensional white disc anymore. It looks like a ball, a gray planet. You can see the mountains and craters clearly. It feels closer than it is. You can see the gas clouds of the Milky Way. You're in the greatest planetarium ever built.

Using our helmet lights to work in the darkness, our first major task was to remove the port solar array and stow it so that we could install the new one. The photovoltaic panels and their metal frame had already been retracted and rolled up, which left this ten-foot pole sticking out of the telescope. We had to fold it up against the body of the telescope. Then Newman would disconnect the old array's connectors from the diode box, the device housed inside the telescope that translates the solar energy into electric power. Then we would remove the array, with each of us grappling it from opposite ends, him at the bottom and me at the top.

To do that, Nancy had to fly me on the robot-arm platform to the top of the telescope at the back of the payload bay. This was tricky for two reasons. With the robot arm, the farther it extends its reach, the more it vibrates and starts to wobble at the far end. It's the same as how your own arm works. When your elbow is bent and your hand is close in to your chest, it's easy to hold something in one place. With your arm outstretched and your elbow straightened out, it's harder to hold that thing still without it vibrating—and I was the thing on the end of the arm being vibrated. The higher I went, the wider the amplitude of the vibration, and the more wobbly it got. I was petrified the foot restraints were going to give and I'd go flying out of there . . . which wasn't possible, but the fear was still real. I was digging my heels into the platform as hard as I could. I was literally thinking: *Feet, don't fail me now.*

The second problem was that the Hubble is forty-three feet tall, which meant I was now standing nearly five stories above the payload bay—and my fear of heights kicked in. I know that sounds crazy for an astronaut. Typically when you're floating above the Earth the distance is so vast that you lose any sense of height, which cancels out the fear of falling. But looking back down at the payload bay, I suddenly had a very real sense of height, like I was hanging off the ledge of a five-story building, wobbling out of control, about to plummet to my death below.

Rationally, I knew that was absurd. Even if my foot restraints came loose—which I knew they wouldn't—the worst that would happen was I'd float there. The rational part of my brain was saying, *Mike, you're weightless. You can't fall.* But the fearful, reptilian part of my brain was screaming, *You're too high up! You're gonna die!* And in the human mind, our rational voice doesn't always prevail. I reached out and grabbed the handrail of the robot arm platform and held on to it with a death grip so I'd be "safe." After that, I could breathe easier. I knew it didn't make any sense. I knew I was being an idiot, but having something in my hand made me feel more secure. I felt better after we got back and other astronauts told me they'd had to do the exact same thing.

Throughout the first night pass and the second day pass we

worked slowly and deliberately. We removed the old array, translated it back down to the payload bay, and stowed it in the carrier for the flight home. As we went to remove the new array from the carrier, we started to enter the second night pass. Now, not only was I facing the most difficult task of the entire EVA, rotating the array into position, I was going to have to do it in the pitch blackness of night. I undid the latches that were holding the array inside the carrier. Then I deployed the mast and engaged the two bolts that would keep it locked in position. Nancy was holding me parallel to the floor of the payload bay as I worked, like I was floating on my stomach a few feet off the ground. When it was time to remove the array, I held on to the frame of the array while Nancy lifted me and it up and out of the payload bay. Once we were clear, she pivoted me 90 degrees upright.

I looked down and in the flight deck windows I could see the faces of my friends watching me intently and quietly rooting for me to succeed. It was the moment of truth. I was twenty feet above the payload bay, on the wobbly end of the robot arm. Other than the light from my helmet illuminating a few feet in front of me, it was completely black all around. Besides the radio, my breathing inside the suit was the only sound I could hear.

I started rotating the array, this massive, king-size mattress thing. Right hand moves an inch. Left hand moves an inch. Right hand moves an inch. Left hand moves an inch. For fifteen minutes, that was all I did. Right hand moves an inch. Left hand moves an inch. Right hand moves an inch. Left hand moves an inch. Every ounce of my energy was focused on that array. A minute or two in, I felt the tiniest wobble. I stopped, held myself perfectly still, and waited. I took only the smallest breaths, inhale, exhale. Finally it was back under control. Then I started again: Right hand moves an inch. Left hand moves an inch. Right hand moves an inch. Left hand moves an inch. Another wobble, another pause, then back to rotating again.

Inch by inch, I rotated the array until finally it was in the proper position. Success! In that moment I felt the sweetest relief. I knew I still had work to do, but I'd faced the biggest challenge of the mission and pulled it off. I looked back down at the payload bay window, and the crew inside were all giving me the thumbs-up. I looked over at the telescope and Newman was smiling and pointing at me and saying, "You're the man!" I was mentally drained. I didn't want to ever do something like this again. Nancy flew me over to the telescope, where Newman and I mated the array's mast with the diode box assembly. That went off without a hitch. Then we unfolded the new array like a book and locked it into place. With that, Hubble had a new power source and a big boost added to its life span.

The rest of the space walk went smoothly. In fact, we were doing so well that we were moving ahead of schedule. At one point we ran into a question we needed Mission Control to answer, about whether or not they wanted us to test a latch on one of the telescope doors. As incredible as it sounds, I'd been out in space for nearly six hours and I'd taken only a few moments to glance at the Earth; I'd been doing my best to ignore it in order to concentrate. Now, while Newman and I waited for an answer, I paused and took a closer look.

It was a night pass. We were over the Pacific, and everything was totally black but for the city lights on Hawaii and a few other islands below. As we came up on California I felt a slight warmth and I knew the next day pass was coming. I looked across the country to see Atlanta all bright and lit up with everyone getting coffee and going to work, but right below me Phoenix and Los Angeles and San Diego were still in complete darkness. I could imagine the tourists at the Grand Canyon, patiently waiting for sunrise to get their perfect golden shot.

The way we experience sunrise on Earth is so gradual. The black outside your window slowly turns to gray. You see a few glimmers of light reflecting off the buildings across the street. In space there is literally a line that bisects the Earth. On one side of it there is darkness. On the other side there is light. The line sweeps west, swiftly and steadily, coming across Europe, across the Atlantic, across Florida, across Texas. I watched this line coming toward me. Then I looked past it up toward the sun. Then I looked back at the line again and realized: *The line isn't moving. The sun isn't moving. We are.*

At that moment I realized that, for my entire life, my perception of reality had been wrong. Every morning you wake up and sit and have your cup of coffee and you watch the sun rise. You don't have any sensation of the Earth moving beneath you. You think you're sitting still as the sun rises in the east and crosses above and sets in the west. But the sun isn't moving. Yes, the sun and the solar system are flying through the galaxy at 45,000 miles per hour. But relative to you and me and the Earth, the sun isn't going anywhere. The whole way we talk about our place in the universe is wrong. "Sunrise" and "sunset" are words that don't make any sense. It's like the song from *Annie*: "The sun'll come out / Tomorrow." No it won't. The Earth will rotate toward the sun tomorrow. That may not be as poetic, but it's reality.

And you know that. Anyone who's taken third-grade science knows that the Earth rotates and revolves around the sun. You understand it intellectually, but you don't *feel* it until you've seen it from in space. *Columbia* had spent the last half hour flying away from the sun, doing a slingshot pass around the Earth, and now our orbit was throwing us back toward the sun again. But the sun had not moved. That's the other thing that hit me: Our sun has been there for a bazillion years, and this has been happening for a long, long time and there's nothing we can do to stop it. People are going to come and go, live and die; bad things are going to happen and good things are going to happen. But nothing we do is going to change this cosmic dance that's been going on since the beginning of time.

My whole life I'd thought of Earth as this place where we're in control of our lives. I'd wake up, go to the grocery store, take my kid to a baseball game. It was this safe, stable cocoon. Now it wasn't that anymore. In space I could see the Earth in relation to the stars and the sun and the moon. The Earth is a planet. It's a spaceship. We're zipping around the universe, hurtling through the chaos of space with asteroids and black holes and everything else, and we think we're safe but, boy, we are right out there in the middle of it.

Once Newman and I finished up and ingressed back into the airlock, I was not the same person I was when I went out. Newman did the depress, the pressure equalized, the inner hatch opened, and before I could get my helmet off, Digger was right there, like he said he'd be, waiting for me to tell him about it. He wouldn't even let me get out of my suit. He was right there in my mug. "What's it like? What's it like?"

"Digger," I said, "you're never going to believe it."

"What?"

"The Earth is a planet."

"*What?*" He looked confused. "Mass, are you okay?"

"It's a *planet*," I said. "It's not what we thought it was back home. It's not this safe cocoon, man. We're out here spinning in all this chaos. The Earth is a planet. The Earth is a *spaceship*, and we're all space travelers."

That's the truth, and that's still how I think of the Earth today.

I walk out my front door in Manhattan in the morning. Everyone else around me sees the street in front of them and the buildings around them and none of that stuff is moving, but they look up and see the sun flying from east to west overhead. I walk down that same street and I know that sun is staying right where it is and it's me and these buildings and this street and this planet that are spinning round and round and hurtling through the void. And the fact that that happens every day, the fact that we exist, is an astonishing thing.

17

MAYBE THIS IS HEAVEN

During the pre-breathe for my first EVA, I nervously eyed the door to space, pondering the mysteries waiting for me on the other side. During the pre-breathe for my second EVA, I fell asleep. Doing a seven-hour space walk is as physically taxing as running a marathon. Here we were doing five of them back to back to back, and on the alternate days you don't get to rest; you're going all day to support the team outside. After a long, long day and a short night's sleep, Newman and I were back in the airlock, waiting to run our second marathon. So I took the opportunity to catch some rest.

My second EVA, installing the Advanced Camera for Surveys was going to be different for a few reasons. Newman would be on the robot arm and I would be free-floating, which was going to give me more opportunity to move around like a spaceman; I was excited about that. It was also my turn to go out first and be the team leader. A motto at NASA is "train the new guy to be your replacement," and that's how we ran the EVAs. Newman was still the lead spacewalker, but he'd shown me how to do everything and this time it was his job to observe me as I took on those responsibilities. After opening the hatch and heading out, I took a few seconds to look around. For that brief moment I was the only human out in space. Anywhere. In the universe. That felt really cool, and I took a moment to let it sink in.

Start to finish, replacing the Faint Object Camera with the Advanced Camera for Surveys took several hours, but we didn't hit any major snags and the installation was a success. Once we knew the ACS was in and powered up, we breathed a big sigh of relief. We were proud of what we'd accomplished. Maybe we hadn't walked on the moon, but that day the crew of STS-109 made a giant leap for mankind. Getting that new camera in was a big deal. Installing the new solar array had been the more difficult task for me physically, but the Advanced Camera for Surveys is one of those things you can point to and say, "This is why astronauts exist. This is why we go to space. This is how we serve the public good." If the ACS did what the engineers said it would do, it was going to unlock the secrets of the universe and help us answer a lot of the big questions about How We Got Here.

There is the one thing EVA training can't do, and that's prepare you to answer big questions like that one. Working in the pool doesn't prepare you for the emotions that can overwhelm you when you're actually out in space. Newman went to fetch the FOC from its temporary stowage location so we could secure it for the ride home, and while I waited for him I took a moment and turned and glanced over my shoulder at the Earth again. As I looked down, the thought that entered my head was *This is something I'm not supposed to see. This is a secret. I'm not supposed to be up here.* I turned my head and tried to go back to my work, but I couldn't help sneaking a second look. I stole a glimpse, and the planet below was so beautiful that I actually started getting emotional. I had to look away.

I was afraid I was going to tear up, and if you get water floating around in your suit, that could be a big problem. There would be a postflight investigation and I would have to admit that I was crying in space. After I'd collected myself, I looked a third time. When I did, the thought that went through my head was *If you were in heaven, this is what you would see. This is the view from heaven.* Then that thought was immediately replaced by another thought. *No, it's even more beautiful than that. This is what heaven must look like—maybe this is heaven.*

I know that might sound strange. There are so many horrible problems here: war, hunger, killing, suffering. But heaven is supposed to be this beautiful, perfect place, and from up there I couldn't imagine anything more beautiful, more perfect than this planet. We might discover life in other solar systems someday, but for now there's nothing but chaos and blackness and desolation for billions of light-years in every direction. Yet here in the middle of all that is this magnificent place, this brilliant blue planet, teeming with life. It really is a paradise. It's fragile. It's beautiful. It's perfection. You have to stop and ask yourself: What in creation could possibly be better than this?

When I flew on that mission, Gabby was eight and Daniel was six. Looking down on the Earth from space, I started thinking about the planet as a father, as a parent. When you have kids, you want to give them everything. You try to find the best house in the best neighborhood. If you can afford to give your kids their own room, you try to fix it up as best you can. You get those blackout curtains so it'll be dark when they need to nap. You put nice toys in there for them to play with. You give them a home. And my thought looking down at the Earth was *Wow. How much God our Father must love us that he gave us this home. He didn't put us on Mars or Venus with nothing but rocks and frozen waste. He gave us paradise and said, "Live here."* It's not easy to wrap your head around the origins and purpose of the universe, but that's the best way I can describe the feelings I had.

Over the course of five servicing missions, only sixteen people spacewalked on Hubble. Just as the twelve Apollo moonwalkers were the only people ever to walk on the moon, the Hubble space-walkers are the only people ever to get out and walk around at that altitude, the only people ever to see the Earth from that vantage point. I was fortunate to experience something that day that only fifteen other people in human history have ever experienced. You can go to the Galápagos Islands or climb Machu Picchu or dive to the bottom of the ocean, but I don't know that there's any experience on Earth that could ever be as extraordinary as being in space. Those first space walks changed my relationship not just with the Earth but with the universe. Forever.

The next morning we executed our fifth and final space walk. The NICMOS cryocooling system installation was a success, and soon the Hubble was ready for redeployment. All satellites experience orbital decay, meaning they gradually slip closer to Earth and eventually burn up in the atmosphere. So once the final space walk was over, Scooter and Digger flew us four miles farther up in order to boost Hubble's orbit and extend its life span.

On Saturday morning, we prepared to say good-bye to the Hubble. The telescope's antennas were remotely redeployed. We maneuvered *Columbia* into a position where the new solar arrays could be exposed to the sun and fully charge up; then a few hours later the umbilical was disconnected and the telescope was switched back to its own power supply. Then, at 4:04 a.m. Houston time, Nancy used the robot arm to lift Hubble from its cradle high above the payload bay. She let go, Scooter slowly backed us off, and the Hubble was on its way. As I watched it grow distant from the window I felt a sense of gratitude and relief. We were sending Hubble off in better shape than when we arrived: mission accomplished.

Sunday was our day off, a day to rest and recuperate, and we needed it. STS-109 had set a new record for space walks on a single shuttle mission. We spent a total of 35 hours 55 minutes, beating the previous record of 35 hours 26 minutes held by STS-61, the first Hubble servicing mission. By Sunday we were exhausted and ready to blow off steam.

Commanders and pilots hate the last couple of days on shuttle missions, because everyone else gets to relax, but they can't because they still have to land safely on Earth. Scooter was going around saying, "We haven't landed yet, guys! We still have to land!" while the rest of us were taking pictures, listening to music, and dancing around in midair. After months of carefully watching every calorie I ate, I was stuffing my face, eating macaroni and cheese like it was going out of style. I took pictures with my Mets jersey and some of my other personal items. That's always fun—you float the object in the air and take a picture of it and the folks back home get a big kick out of it. We took our crew photo. We took turns doing private video conferences with our families. I did some tricks for them, like eating floating M&M's and doing somersaults in the air.

I spent most of my day off on the flight deck, floating at the window, listening to music, and staring out into space. When we first got to orbit I was obsessed with looking at the Earth during the day, seeing the Himalayas and the Sahara Desert and all these amazing formations from 350 miles up. Out at the edge of the planet, you can see the line where our atmosphere meets the stars, and it has this bluish-greenish hue to it that's absolutely beautiful.

By the end of the mission I'd grown to enjoy the night passes more. You'd see shooting stars, meteors burning up in the atmosphere below. You'd see fishing trawlers lit up off the coast of Japan. But the lights of the cities are the main thing. You're looking at them through the atmosphere so they have this diffuse, orangey glow to them. You'd look for the different patterns in them. Los Angeles sprawled out. New York burning like a jewel. At night, even compared to other developed countries, the whole United States is lit up like a Christmas tree, especially along the coasts. Cuba and North Korea are total blackouts.

Lightning storms at night are amazing, too. You'd be over the ocean looking down on total blackness. Then lightning would flash, illuminating the features of the clouds from within. You'd see a flash, and then another flash, then another. There would be three or four of them in a row. Then a lull. Then three or four more. It was like a form of communication, like a sequence, like the clouds are alien creatures speaking to each other in code.

The music is a key part of it. You have to have the right sound track for space. My first flight we had CDs and Discmans. My second flight we had iPods. That was better. Some music works during the day pass, some works better at night. Sting, Phil Collins, Coldplay, and U2 are fantastic during the day. Radiohead is perfect at night. It's possible that Radiohead's *OK Computer* was recorded specifically to be listened to in space, and that everyone who's heard it on Earth is missing the full experience. Movie soundtracks work well at night, too. John Barry's *Dances with Wolves* soundtrack. Thomas Newman's *Meet Joe Black* soundtrack. I listened to those over and over again, floating at that window, watching Earth spin below.

It was good to have some time to myself to unwind. That first flight was incredible, but looking back on it I have to say it was more intense than enjoyable. Our sleep schedule was way off, waking up at ten at night and going to bed at two in the afternoon. We were always working to catch up. I recognized the feeling from Cold Lake. It was poor expedition behavior. Everyone was exhausted and stressed out. Weirdly, after thirty-nine years of trying to get to space, part of me wanted to hurry up and get home. I wanted to get the first flight out of the way so that I could come back again as a veteran. At the time, astronauts were getting five and six flights each. I said to myself, *This was just to get my feet wet. I'll be back.* I sort of discounted the experience in my mind. I took it for granted. It never occurred to me that anything would change.

Monday morning we woke up and started getting ready to go home. I wasn't particularly concerned about reentry. The biggest worry in spaceflight is launch. Rockets blow up on ascent. On Hubble, our second biggest worry, from a safety standpoint, was the EVAs. We were concerned about losing somebody on a space walk. That same vigilance didn't apply to reentry. It's not that we were blasé or complacent about it. We knew that it posed serious risks, but we felt like we'd mastered it. The Russians had lost some people on reentry with the Soyuz, but we never had. Going all the way back to Alan Shepard's first Mercury flight, no American astronaut and no American spacecraft had ever been lost coming back from orbit. The fear that gripped me on the launchpad going up—I didn't have that coming down. It never occurred to me that Scooter and Digger wouldn't get us home in one piece.

Returning from space, the shuttle normally hits the Earth's atmosphere at Mach 25, producing enormous amounts of heat and friction. Coming back from Hubble, because we were higher, we actually hit Mach 26, which means that the Hubble astronauts have flown the highest and fastest of any astronauts in the shuttle era. Since I was on the mid-deck, I couldn't see anything. I could feel it warm up a bit, but nothing unbearable. I had a bag of M&M's tethered to my seat. It was floating above me and then suddenly it fell to my lap. I felt heavy. I felt my body being pushed down in my seat, my arms and legs having weight again.

Minutes later, Scooter and Digger brought us in for a perfect landing at Kennedy Space Center at 4:42 a.m., capping off a ten-day, twenty-two-hour and ten-minute mission that covered 3,941,705 miles. Once we came to a stop, the ground crew came in and helped us out of our harnesses. When you first stand up, not only are you weak and wobbly from being in space for almost eleven days, but your spine is crunching back down to its normal height and the sensory inputs from your inner ear that weren't there before come rushing back into your brain. You have to stand up slowly, and you feel like you're going to fall flat on your face. For a while you're walking around like Herman Munster, trying to get your bearings. We hugged and took some photos and then went inside to the crew quarters, where our families were waiting.

The single best thing about coming back to Earth was seeing my kids again. As soon as I walked in, they ran over and gave me big hugs. The flight surgeon had stressed to us that we weren't supposed to pick them up, since our bodies were still adjusting to gravity, but I couldn't resist. I grabbed them up and held them as tight as I could. Daniel's Little League season was about to start. I was going to be a coach, and I couldn't wait to hit the field with him. I was so grateful to be home alive.

After seeing my family, I went back to crew quarters to get changed into my civilian clothes. My room was exactly as I'd left it. Everything was the same, but I was not the same. So much had happened in between. It was the first time I had really been alone in nearly two weeks. That whole time I had been around my crewmates in close quarters, staying focused on the mission, keeping my emotions in check. But there, alone in my room, I started thinking about the journey I had taken, the incredible beauty I had seen. I started to cry uncontrollably. They weren't tears of sadness or even happiness, really. I was overwhelmed. It was a release of all these different emotions I'd been keeping pent up inside: the joy and the exhilaration and feelings of childlike wonder. I sat there for ten, fifteen minutes and cried and cried and let it all out. Then I pulled myself together, took a real shower for the first time in weeks, put on my jeans, and re-entered the Earth.

Out on the tarmac, the ground crew was already hard at work, taking the *Columbia* orbiter through its postflight inspection. Soon our spaceship would go back into the Orbiter Processing Facility to be readied for its next flight, STS-107. Shuttle missions are numbered in the order they're assigned, not in the order they fly, which is why we went first despite having the higher flight number, 109. What had happened was this: The crew for 107 was assigned about six months before we were. Theirs was a routine science mission, doing experiments in the Spacehab research module in the shuttle payload bay. *Columbia* needed to be overhauled, and there were some issues that forced 107 to delay. After that initial bump, they kept getting pushed. The experiments they were doing weren't time critical, but the station assembly flights were. So those missions kept taking precedence, pushing 107 further and further back until eventually they ran up against us, the next crew assigned to fly on *Columbia*. Since Hubble was the higher priority, management looked at the flights and said, "Okay, let's swap 'em," and 107 got bumped again. NASA flipped us in the flight order. We got their spot. They got ours. We came back. They didn't.

part

5

Russian Roulette

18

THE STORY OF SPACE

There's a great scene in *The Right Stuff* where Chuck Yeager and some of the Mercury astronauts, Gus Grissom, "Gordo" Cooper, and Deke Slayton, are drinking at Pancho Barnes's Happy Bottom Riding Club, the bar in the Mojave Desert where the test pilots hang out. This Air Force liaison is trying to tell them about the need to get good press. Yeager blows him off. He doesn't have time for reporters. "Them little root weevils that crawl around poppin' off cameras in your face," he calls them. But the Air Force guy has a point: Good press drives public opinion, public opinion drives public policy, and public policy decides who gets the money to fly. He says, "You know what really makes your rocket ships go up? Funding. That's what makes your ships go up. No bucks, no Buck Rogers."

Everyone in the early days of the space program learned that lesson well. The president and NASA and the astronauts did an incredible job of selling the space program to the American people. Kennedy's address announcing the Apollo program was one of the great presidential speeches of all time. He challenged us. He excited us. We reach for impossible things, he said, "not because they are easy, but because they are hard." He told us the moon shot would be "the most hazardous and dangerous and greatest adventure on which man has ever embarked." The space race had everything: good guys and bad guys, us against the Soviets, John Glenn versus Yuri Gagarin. It was Neil Armstrong and Buzz Aldrin in an all-out dash to the moon, a story so compelling that it captured the whole world's attention.

After Apollo, everything changed. We'd won the race. It was time to start a new chapter to get everyone excited again. NASA's original plan was ambitious: a reusable space plane, a space station in low-Earth orbit, outposts on the lunar surface, a manned mission to Mars. But the support from the public wasn't there; people thought the story was over, and they changed the channel. A recession hit, budget cuts followed, and the big-spending days of the space race were over.

The reusable space plane, the shuttle, was the only part of the plan that survived. It was conceived as a spacecraft capable of taking crews and cargo to the different space stations between here and the moon. Once those projects were scrapped, we still needed a reason to build it—instead of designing a vehicle to serve a purpose, NASA had to come up with a purpose for the vehicle it had designed. In the end, the shuttle was sold as a one-size-fits-all vehicle that would do a bit of everything: It was a flying space lab, a cargo ship, a way to launch and service satellites. After the excitement of the space race, the shuttle was sold on the premise that it wasn't going to be exciting at all. It was routine, everyday access to low-Earth orbit, no different from an eighteen-wheeler hauling freight from Indianapolis to Buffalo.

As the budgets got smaller, the story got smaller, too. Fewer bucks, fewer Buck Rogers. With astronauts today, people respect the title, but they don't know who we are. The Mercury and Apollo astronauts were true celebrities. They defined their era. They were the epitome of cool. In the shuttle era, the culture changed. At NASA today, there's no such thing as individual accomplishment— there is no "I" in "team." There's a pecking order to everything. The commander is the spokesperson for the crew, and everyone defers to him or her. Any individual attention you get, you're expected to deflect it back toward the crew and the mission and NASA itself. It's "we" succeed and "we" fail, and that's that.

On the one hand, it has to be that way. When you're in space, it's life and death and everybody's counting on everybody else. It can't be about any one person's ego. You can't have a crew where one person thinks he's LeBron James and the other astronauts are only there to support him. On the other hand, heroes make a good story. The way that NASA started telling the story of space after Apollo, we made the mistake of thinking that the public was invested in the mission, in the objectives of science and space travel. That's not really true. There is a very dedicated community of people who care about "space." They get excited about the discovery of black holes and throw themselves into debates over whether Pluto is a planet or not. The general public doesn't care that much about space—they care about people in space because they can identify with them.

The Soviets actually landed on the moon before we did. They had unmanned probes up there years before Neil Armstrong and Buzz Aldrin showed up. Does anyone remember? Does anyone care? No. America won the race to the moon because America put people on the moon. People are fascinated by other people. Humans like to watch other humans doing awesome things because it makes us feel awesome about being human. Going to space is awesome. People don't come to watch the launches at Cape Canaveral because NASA creates jobs. They come because they want to hear the rockets roar and feel the ground shake and watch the night sky turn into day.

When I watched the moon landing on TV and went to see *The Right Stuff* in the movie theater, I didn't dream about going to space—I dreamed about being an astronaut, because they were my heroes, because their lives looked awesome and exciting and fun. That's what inspired me. And when I finally made it to NASA, I wasn't disappointed. I was flying around the country in a T-38, hanging out with living legends like John Young and Alan Bean. I was living *The Right Stuff* every day, but I wasn't seeing that reflected back to me in the stories being told in the news and in popular culture.

There's a famous episode of *The Simpsons*, "Deep Space Homer," in which Homer goes up on the shuttle. It starts with him watching a launch on TV and an announcer is describing the crew. "They're a colorful bunch," he says. "They've been dubbed 'the Three Musketeers.' There's a mathematician, a different *kind* of mathematician, and a statistician." The joke is that astronauts have become so boring, the public cares so little, and the Nielsen ratings for shuttle launches have fallen so low that NASA is in danger of losing its funding forever. That tells you a lot about the way NASA was being perceived, but nothing could be further from the truth. The people I work with are genuinely funny, colorful, smart, dedicated, brave people—a bunch of eccentric genius PhDs and gung-ho *Top Gun*–type military test pilots. We love what we do and we get to spend our days doing incredible things that shape the future of all humankind—and we're having a blast doing it. But if the public doesn't know that about us, that's our fault.

I remember watching NASA TV one time. NASA TV is way

down on your cable channel guide, if it's there at all. It provides live coverage of launches, space walks, and other events. I turned it on, and this buddy of mine, "CJ" Sturckow, was being interviewed. CJ is a Marine pilot who flew forty missions in Desert Storm. He used to pit-crew for off-road truck races down in Baja. He's hilarious, one of biggest characters you'll ever meet. But this interview was the most boring interview I'd ever seen in my life. The questions were about dry, technical stuff. They weren't letting any of CJ's personality through. I was, like, *Where is my friend? Who is this undertaker?* The lighting was bad. Everything about it was bad. The result is that nobody's paying attention and the American public doesn't know why we do what we do anymore.

As an ASCAN, we did receive some media training: how to talk to reporters, how to give a speech and a presentation. Bill Wallisch is a public-speaking instructor who's done the media training at the Johnson Space Center for years. He was our coach. For my training presentation, I decided to do something personal. I told the story about my dream of becoming an astronaut. I showed the picture of me as a kid in my spaceman costume with my Astronaut Snoopy. I showed pictures of Columbia and MIT and talked about the route I took and how hard I worked and how I never gave up. I threw in a few jokes about myself. Bill liked my presentation so much, he took the videotape of it and started using it to train the classes who came in after me.

I might have been a hit in media training class, but for a while that didn't mean a whole lot; ASCANs and rookies typically don't engage with the media. I did little things. A few TV interviews. I spoke at my niece and nephew's school in Vermont. My first chance to step out came with the postflight awards ceremony for STS-109. After every shuttle landing there's an awards ceremony where we get our flight medals and then stand onstage to take questions.

Most of the ceremony was routine. I remember one of the questions was "Do you ever think you'll send kids into space?" But to be honest I wasn't really paying attention. The rest of the crew did most of the talking; they were the veterans and Scooter was the commander. I was just the rookie. I figured nobody was going to ask me anything. I was mostly daydreaming and thinking about where we were going to eat afterward. Then I heard, "Mike? . . . Mike?"

I looked over and saw Scooter waving at me. I said, "What?"

"Your story? About seeing the planet?"

"What about it?"

"Can you please *tell* it?"

I hadn't heard what the question was, but I obliged and launched into my story about watching the Earth move and telling Digger in the airlock and how awesome it was and how my relationship to the universe changed and the whole thing. Everyone applauded. They really responded to it. Sean O'Keefe, the NASA administrator, wrapped up by saying, "Somebody asked if we're going to send kids into space. We have sent kids into space. We sent Mike Massimino."

I didn't think my story was anything special at the time, but it was a bit of humanity injected into a press conference that had mostly been about launch trajectories and repair schedules. I figured the average person doesn't want to hear that. We just flew in *space* on a *spaceship*. What's it like out there? What's our place in the universe? What's it like to achieve your childhood dream? That's what people want to know.

O'Keefe loved it. He had just joined NASA—he had been appointed by President George W. Bush—and STS-109 was the first flight on his watch, so he was there for our launch and for our landing. After the ceremony, he came over to me and said, "That was a fantastic story. We're getting you on the circuit." From that day forward, I was his guy. I started getting assignments direct from his office. Whenever NASA needed a public face somewhere, O'Keefe would say, "Send Massimino." I started speaking at different conferences and events. I did a public service announcement. *Apollo 13* premiered that year on IMAX, and I was sent to represent the agency and talk to reporters and take pictures with Jim Lovell and Ron Howard.

Traveling back and forth to these events, I had a realization. When it came to the work of being an astronaut, I was good but I wasn't the best. I wasn't the best guy with the shuttle systems. I wasn't even the best spacewalker. But maybe I could be one of the best at telling the story of space. Maybe I could make it fun and lively and adventurous like it used to be. I had an interest in and a talent for communicating with people and helping them understand what our jobs are like. I didn't want to go out and be self-promoting, but I wanted to help promote the other people in the program and the program itself.

While O'Keefe was busy sending me out, my baseball obsession turned into a massive public relations event of its own. It started months before my flight when I ran into Ellen Baker, one of my fellow astronauts, in the gym. Ellen is from Queens and her mother, Claire Shulman, had been Queens borough president a few years before. Ellen knew I was a huge Mets fan and asked if I'd thought of flying anything from the team. I said sure, but I didn't know how to get in touch with them. So she reached out to her mom, who reached out to the Mets' owner, Fred Wilpon. That same day Ellen sent me a note: The owner of the New York Mets is expecting your call. Here's his number.

Getting the owner of a Major League Baseball team on the phone in less than a day—that was one of those moments when

I realized how much leverage astronauts could have in getting the public's attention if we used it more. I called Wilpon and he put me in touch with the team's media affairs officer, Jay Horwitz. I asked if I could fly John Franco's jersey. Franco was a relief pitcher. Team captain. A hometown guy. He grew up in Bensonhurst, Brooklyn. My father worked for the fire department. His father worked for the sanitation department; he wore an SDNY T-shirt under his jersey. Jay sent me a signed Franco jersey. I flew it on the shuttle and took pictures wearing it in space.

That April, the Mets were in Houston for a series with the Astros, and Jay invited me to come out on the field and meet the players before the game. I remember being out on the field at Minute Maid Park and Jay told me that Bobby Valentine, the Mets' manager, wanted to meet me. Valentine walked over with this big smile, grabbed me by the shoulder, and said, "I love what you do! What you guys do is the greatest thing in the world! What a pleasure to meet you. I've got a favor to ask."

"What's that?"

"Can I have your autograph?"

He already had a print of my official NASA photo ready for me to sign; he said he wanted to put it up in his restaurant. I had to stop for a second. This was the manager of the New York Mets, and *he* was asking *me* for an autograph. My whole world had shifted 180 degrees. How many times had I been in the stands, one of thousands of people holding out a ball, trying to get a player to sign it, with this curtain between me and my heroes? Deep down I was still the same kid who idolized these guys, but now that curtain had lifted and I was on the other side. Maybe NASA wasn't as exciting to people as it used to be, but there was still a lot of love for the space program out there.

I signed Bobby's picture for him, and then he and Jay hit me with their big idea. They wanted me to come to Shea Stadium and present the flown jersey to the team by throwing out the first pitch at a game. Now my world had really shifted. I'd dreamed of standing on that pitcher's mound more times than I'd dreamed of walking on the moon. I started talking to the team about what games would work and we went back and forth and the only day that fit everyone's schedule was June 15: a Saturday day game in a Mets-Yankees series, one of the biggest rivalries in baseball.

When I walked into Shea that Saturday it was like walking into the Roman Colosseum. MLB Game of the Week. Sold-out stadium. Fifty thousand Mets fans screaming "Screw you, Yankees!" and plenty of worse things laced with profanity I had to explain to my nine-year-old daughter and six-year-old son.

Gabby and Daniel were going to go out on the field with me and hold the jersey for the crowd to see while I threw the pitch. I also invited my uncle Romeo, who used to go to the games with me and my dad and my cousin Paul. Franco was going to catch my pitch. He and I talked a bit about what I was going to throw. My childhood buddy Mike Q and I had been practicing all week. He actually went and dug out his old catcher's gear from high school and we went to the park out in Franklin Square. I threw pitch after pitch after pitch, the same way I used to throw the ball against the steps of my house over and over and over, pretending I was pitching at Shea—and now I was.

Right before the game started, Bobby Valentine came over to say hi. I asked him if he had any advice. "Two things," he said. "First, throw it higher than you think you need to. Johnny's a professional baseball player. If you throw it anywhere in the air, he will catch it and make it look like a strike. If you bounce it, he cannot help you. Aim it high."

"Okay," I said. "And the second thing?"

"Yeah, your pants are filthy. You been playing in the dirt? Clean 'em off."

I *had* been playing in the dirt, writing "Go Mets!" in the ground in front of the dugout with my kids. I brushed myself off and got ready to go out. The announcer came over the PA as we walked out: "Astronaut Mike Massimino is a lifelong Mets fan from Franklin Square, New York, who just flew John Franco's jersey on the space shuttle!" The fans went insane, screaming and yelling. It was like, *Take* that, *Yankees. You think you're so great, but our guy's been to* space*!*

I walked out and stopped at the mound. As I looked around I saw the camera out in center field and I started getting choked up. How many times had I seen the view from that camera as a kid and dreamed about doing this? Franco was looking at me from home plate. He could tell I was having a moment, and he gave me a big smile.

I stepped up on the mound. Franco got down in the crouch and was pounding his mitt. I tuned out all the noise. I couldn't hear a thing. John Franco was the only other person in that stadium. The focus and concentration I had rotating that solar array in space? This was that intense. My mind was racing: *Higher than you need to . . . Higher than you need to . . . Massimino, you are NOT bouncing this pitch.* I went into my windup, let go, and *wham*! Right in his glove. From the mound I could tell it was a bit low, but Franco snapped it up and tucked it in and made it look like a strike. The second he caught the pitch it was like somebody flipped the sound back on and I could hear the whole crowd roar. I realized they'd been going crazy the whole time.

Which was a better moment: Seeing the Earth from 350 miles in space or throwing a pitch at Shea Stadium? I honestly don't think I could pick. From the age of seven, other than my dad, my two biggest heroes were Neil Armstrong and Tom Seaver. Those were the guys I wanted to be, and here I was, living both of my childhood dreams at the same time. How lucky was I? In my mind, I imagined that maybe there was another seven-year-old boy or girl out in the stadium that day. Maybe it was a formative moment in their lives and they saw me throw that pitch and thought, *Wow, there's a job where you get to fly in space* and *walk out on a major-league ball field? Being an astronaut is pretty cool.*

That same trip home I went back and visited my elementary school and my high school. They threw a parade in my honor. It was so gratifying to see kids light up when I told them about being weightless or what the stars look like in space. From that point forward, other than actually being in space, sharing the story of space became one of my favorite things to do. I wanted to inspire young people to follow their dreams the way my heroes inspired me. O'Keefe kept sending me out on more press assignments, and I was always happy to go and talk up the space program and get people invested in seeing our journey as a part of their journey.

That September was the first anniversary of the 9/11 attacks. All of the different government agencies were doing something in observance. Along with 60 police officers, 343 firefighters died when the Twin Towers collapsed. O'Keefe knew about my connection to the New York Fire Department through my father, and he asked me to come up and speak to the staff at NASA headquarters for the occasion. I didn't go up with a script or anything. I just shared some of my personal experiences. I talked about what the fire department meant to me growing up, how my father had taught me about the value of public service. Astronauts and ball players may have been my idols, but because of my dad, I always knew firefighters as the heroes next door, the everyday people who risk their lives to keep us safe.

Thousands of people died on 9/11, but we especially remember the cops and the firefighters because they died serving the public good. They ran into the towers when everyone else was running out, and that makes them exceptional. It doesn't matter if it's a first responder charging into a burning building, a man walking on the moon, or two ballplayers locked in a grudge match. People identify with other people. We celebrate them in their moments of triumph, and we mourn them when tragedy strikes.

19

FEBRUARY 1, 2003

Not long after Rick Husband and I served as family escorts together on STS-87, we were at a dinner in honor of Leonid Kad-
enyuk, a Ukrainian astronaut who was flying on the shuttle. The
subject of family escorting came up. Neither Rick nor I had flown
at that point, but Rick turned to me and said, "If I ever fly in space,
the person I want to be my family escort is Mike Massimino." I told
him it would be my honor. A few months later, Rick was assigned
to STS-96 as a pilot. Kent Rominger—we called him "Rommel"—
was the commander. They asked me to be one of the escorts, and
I immediately said yes. After everything Rick did for me when my
father was ill, it was an honor for me to help his wife, Evelyn, and
their two wonderful kids, Laura and Matthew, deal with the dif-
ficulties of sending a loved one off on a space flight.

Everything on STS-96 went great. I enjoyed being a family escort. It was, for me, one of the most important things I was able to
do as an astronaut. Part of it is just fun. There's lots of parties. The
children are excited about watching their moms and dads going

into space. You get to entertain them and see the excitement of the launch through their eyes. It's also a serious responsibility. You handle the logistics, getting the family checked into the hotel and arranging their rental car, but your main job is to be there in case something goes wrong. The morning of the launch, you're responsible for going to the hotel and picking up the spouses and the kids. The protocol is that they have to have their bags packed before they can leave. They'll be coming back to spend another night in that same hotel, but their bags have to be packed and ready to go, because if there's an accident, those families aren't going back to the hotel to pack. They'll be rushed onto a plane and taken back to Houston and somebody's going to be assigned to collect the bags for them. It's an eerie moment. You pick them up and you say, "Where's everybody's bag? Everybody set?" What you're actually saying is "Make sure you're packed in case your spouse blows up in front of your children today." When someone asks you to serve as their family escort, they're entrusting you with a very serious responsibility.

When Rick Husband was assigned to be commander of STS-107 on *Columbia*, he asked me to serve as a family escort again. At the time I was going through CAPCOM training, which was taking up a lot of my time. I checked with Mark Polansky, the head of the CAPCOM office, and was told I couldn't do it. I needed to be in Houston and couldn't be going down to the Cape. I was upset. Rick wanted me to be there for his wife and his kids, and I felt like that should have been the priority.

I was close with several of the other people on that flight, too, and I wanted to support their mission. Mike Anderson was probably the crew member I knew the least. Like Rick, he was an Air Force pilot, also very religious, a super-nice guy. I flew with him a few times as a backseater in his T-38. Dave Brown was a Navy pilot and flight surgeon. Dave was a bachelor, so I didn't see him around as much as I did the astronauts with kids. But he and I happened to be down at the Cape together in 1997 when they were having the wrap party for *Armageddon*, so together we got to live the life of Hollywood decadence for one night, feasting on lobster and king crab legs. Kalpana Chawla was the first Indian-American woman astronaut. We called her KC. She had been the astronaut office robotics liaison when I was at McDonnell Douglas, and we'd worked closely on developing the robot-arm display. She was one of the most brilliant astronauts I ever worked with and a pleasure to be around as a person.

Willie McCool, 107's pilot, had the greatest name of any astronaut ever. He was a Navy pilot who launched off aircraft carriers in fighter jets in the middle of the night, but he also wore clogs and wrote poetry to his wife. He was like a granola, hippie fighter pilot. One day when Daniel was up at the office with me, Willie put him in an office chair and flew him around the hallways at top speed playing fighter pilot. Daniel loved it. Laurel Clark was a Navy flight surgeon and a rescue diver. Her nickname was "Floral" because she wore bright, flowery clothes and she had a bright personality to go along with it. Laurel and I were in the same class and lived in the same neighborhood. She and her son, who's about Daniel's age, moved to Houston a few months ahead of her husband, so we ended up doing a lot of activities together with our kids.

Other than Rick Husband, the crew member I was probably closest to was Ilan Ramon. After his reelection in 1996, President Clinton had announced that the first Israeli astronaut would be flying on the shuttle as a part of the special relationship between our two countries. That was Ilan. He wasn't a part of any astronaut class, but we made him an honorary member of the class of '98. There were thirty of them that year, coming in right after our class of forty-four, which meant they were going to wait a long time to fly. We named them the Penguins—the flightless birds. They'll fly when Houston freezes over.

Many international astronauts leave their families at home, come to Houston for a couple of years, fly their mission, and then leave. Not Ilan. He wanted the whole experience. He moved his family over: his wife, Rona; his three young boys, Asaf, Tal, and David; and his daughter, Noa. Given the arrangement that Clinton had made with Israel, Ilan was going to fly no matter what. He wasn't waiting in line with the rest of the Penguins, but he didn't act that way at all. He insisted on being treated like everyone else. He didn't want to be a person who flew on the space shuttle one time. He wanted to be an astronaut.

Carola and I got to know Ilan and Rona through Steve Mac-Lean, a Canadian astronaut. There was a group of astronauts who wound up socializing because our wives went cycling together. Willie McCool and his family were a part of that group, too. We would have weekend barbecues and spend holidays together. Willie would organize huge ultimate Frisbee games in the park, fifty people out on the field, parents and kids, playing together. It was a blast.

Ilan was an exceptional person. He was an F-16 pilot in the Israeli Air Force, a colonel. He graduated at the top of his class in flight school and was a decorated veteran of the Yom Kippur War. He also took part in one of the most important military operations in Israel's history, Operation Opera, a preemptive strike to destroy a nuclear reactor that Saddam Hussein was building in southern Iraq in 1981. A flight of F-16s armed with heavy explosives flew out in an early dawn raid to wipe out the target. Ilan was the last pilot in the formation, meaning he was the most likely to take antiaircraft fire and get killed—the most dangerous spot in an incredibly dangerous mission. I didn't know any of this about Ilan at the time.

He was Israel's Chuck Yeager, their John Glenn, and their Neil Armstrong, all rolled into one. But I never would have guessed it from talking to him. He was kind and unassuming and respectful to everyone.

A few weeks before STS-107 flew, Steve McClain had a group of us over for a Christmas Eve party. Ilan and Rona were there. He was amped up and excited about his mission. We made a few jokes about getting swapped in the flight order. "You cut us in line!" That sort of thing. It was a fantastic night. Shortly after that the crew went into quarantine. Two days before the launch I ran into Evelyn Husband at the YMCA. I told her how bummed I was that I didn't get to be their escort, but if she needed anything to let me know. Then on January 16 they flew.

On January 31, their last day in space, I was doing a CAPCOM shift for the astronauts on the space station, Don Pettit and Ken Bowersox, who were up there with the Russian cosmonaut Nikolai Budarin. I got a call late in the afternoon from the flight surgeon, Smith Johnston. The last thing a shuttle crew does the night before they come back is a private medical conference with the flight surgeon to make sure everyone is in good shape for reentry. Smith liked to bring on a "mystery guest" at the end of his calls to say hi and wish everyone well for the ride home. Small things like that are important for making you feel connected while you're off the planet. Smith asked me if I would be his mystery guest. Carola and Gabby were away on a camping trip with her Girl Scout troop, and I had to get home that night to help Daniel finish his car for the Cub Scouts' Pinewood Derby, which was happening first thing in the morning, but I said of course I'd be happy to do anything I could for the crew. Smith asked me to stop by around five o'clock, right about the time they would be in their pajamas getting ready for bed.

The flight surgeon does his calls in a private room right off Mission Control. Once Smith had finished the medical updates, he invited me in. The Ku antenna on the shuttle was already stowed, so it was a voice call, not video. The crew had to guess who I was. Smith made it easy. He introduced me by saying, "He's from New York and he's very tall." Laurel Clark shouted the answer: "Mike Massimino!" I could hear them laughing and cheering in the background. "How is it up there?" I asked. Laurel joked, "Oh, Mass. It's not good. I don't think you should ever come back. I'll take your spot next time."

I spoke mostly to Rick. That's how it is when you call the shuttle: You talk through the commander. "We really appreciate you taking time out of your day to come talk to us," Rick said. "Thanks so much. It's great to talk to you, and we look forward to seeing you when we get back." That was Rick, grateful and generous. We talked for about ten minutes, chatting back and forth. I could tell they were happy. They'd had a great flight and couldn't wait to get home. We said good night and I wished them luck. Smith signed off, and that was that. I ducked out to go pick up Daniel, totally unaware of the fact that—along with the CAPCOM who'd be talking to Rick and Willie during reentry—Smith Johnston and I were the last people on Earth who would ever speak with them.

Daniel's Pinewood Derby race was at 8:00 a.m. up at his school. His car was designed like a Blue Angel F-18 with wings on the back and Blue Angel stickers on it. Daniel got dressed in his Cub Scout uniform. Since I was an assistant troop leader that year, I was wearing my uniform, too. We drove over to the school. In the parking lot as we were walking in we ran into one of the other fathers, Mike Lloyd. Mike was an Army reservist, an enlisted guy. He had a Walkman or some kind of portable radio he was listening to. He saw me and said, "Have you heard anything?"

I said, "Heard what?"

He looked down at Daniel. "Daniel, why don't you go on inside."

I told Daniel to go in and that I'd meet him indoors. Once he was gone Mike said, "It looks like the shuttle came apart in the sky. They're reporting debris over East Texas."

I stared at him for a moment, dumbstruck. Then I ran and found a phone and called Steve Smith. He answered, "Hey, Mike! What's up?"

I said, "Turn on CNN and tell me what you see."

He went and turned his television on. I could hear the muffled sound of the TV, but Steve wasn't saying a word.

"Steve, what do you see?"

"It's an accident. . . . It's bad."

"What do we do?"

"Go to the office," he said. "Now."

I ran and found Daniel and arranged for one of his friends' moms to look after him and get him home. Then I left and raced to the office as fast as I could. I felt completely helpless. I felt like I should be in Florida with the families. I needed to get to the office to hear what was going on, but at the same time I was dreading it. I wanted to wake up and find that the whole morning had been a nightmare.

As I was pulling in, other astronauts were arriving, too. I remember seeing Lee Morin. He was wearing a suit—8:00 a.m. on a Saturday, and he had on a black suit. Because he already knew. Lee was Dave Brown's casualty assistance care officer, or CACO. If you get killed, your CACO is the person who handles all the arrangements for your family and helps your spouse with whatever he or she needs. I was still in shock. But Lee was a Navy guy. He knew. He was ready.

When I walked in I saw Kevin Kregel in the hallway. He was standing there shaking his head. He looked up and saw me. "You know," he said, "we're all just playing Russian roulette, and you have to be grateful you weren't the one who got the bullet." I immediately thought about the two *Columbia* missions getting switched in the flight order, how it could have been us coming home that day. He was right. There was this tremendous grief and sadness, this devastated look on the faces of everyone who walked in. We'd lost seven members of our family. But underneath that sadness there was a definite, and uncomfortable, sense of relief. That sounds perverse to say, but for some of us it's the way it was. Space travel is dangerous. People die. It had been seventeen years since *Challenger*. We lost *Apollo 1* on the launchpad nineteen years before that. It was time for something to happen and, like Kevin said, you were grateful that your number hadn't come up.

We gathered in the sixth-floor conference room, surrounded by the commemorative plaques of all the missions that had gone before. It was packed. Everyone had been called in. There wasn't a whole lot of wailing and crying. It was more quiet, somber: people sitting around in shock with blank, thousand-yard stares, trying to process what had happened. A few of our leaders got up and spoke. Ellen Ochoa, head of flight crew operations at the time. Kent Rominger, who was now head of the astronaut office. I mostly remember former astronaut Bob Cabana, who was deputy chief of JSC at the time, getting up to speak. He confirmed what we already knew, that the crew had been lost. He said, "This is a terrible day for the astronaut office. This is going to be one of the worst days that any of us will ever have in our lives. But we've got to get through it. The families are on their way back from Florida. As devastated as we are, just imagine how they're doing. The first order of business is to take care of them."

When the meeting broke, it was chaotic. We had protocols and contingencies in place to deal with the loss of a shuttle, but the truth is, no one really knew what to do. You did what you thought was needed. Some people were already talking about the investigation and volunteering to help with that. I knew that I needed to be helping with the families. I was still upset that I wasn't in Florida right then. I should have been there like Rick had asked me to be. Andy Thomas, deputy chief of the astronaut office, was coordinating the family support. Even with the escorts and CACOs already assigned to the crew, they were going to need extra help. I went to Andy and told him I was available. Steve Lindsey was Rick's CACO, and Andy more or less told me to go with Steve and do whatever was needed. The families were scheduled to arrive at Ellington from the Cape at 3:00 p.m., and it was decided we would meet there.

I called Carola. She and Gabby were cutting their camping trip short and coming home. I asked her if she needed anything before I disappeared for the rest of the day. She said we needed milk. Life doesn't stop, even at moments like this one, so I went by the store to pick some up. This was around one in the afternoon. I was in a fog. I was a zombie. I hadn't eaten anything all day. I got the milk and a few other things and I was standing in line to buy them when I completely zoned out. The next thing I remember is the groceries being bagged up and the cashier saying, "Sir? Will that be cash or credit? Sir? Sir? Hello?!" I had my wallet out, but I'd forgotten why I was in the store. I just stood there staring into space. One of our neighbors, Theresa Harrigan, was behind me in line. She reached over and took my wallet and paid the cashier for me. She said, "Be nice to this man. He's had a rough day." I thanked her and took the groceries and left. It's funny what you remember.

I dropped the milk at home and drove to the airfield. The

CACOs and a bunch of other astronauts were there. When the plane landed from the Cape, we helped everyone get their luggage and pack up their cars. Rick's wife, Evelyn, was in the passenger seat of a minivan. She put down her window and I told her I was sorry. She took my hand and said, "You know how much Rick loved you. He loved you so much." It was like *she* was the one comforting *me*. But that's how Rick and Evelyn were.

At 1:00 p.m., NASA made a public statement confirming the total loss of *Columbia* and its crew. An hour later President Bush gave a speech from the White House. At 3:20 p.m. it was announced that all future shuttle flights were suspended pending the accident investigation. We knew that would be the case, but now it was official. Don Pettit and the other astronauts on the space station were, at least for the time being, stranded. I was Don Pettit's CACO. Don and Willie McCool were tight. Don had invented a chess board with Velcro pieces that you could play in space, and he and Willie had had a game going, e-mailing their moves back and forth to one another.

After I left Ellington, I went over to Don's house to spend a couple hours with his wife, Mickey, and their twin boys, who were only about a year old. Her husband was 250 miles up in space, and his ride home had just been canceled. We knew we could get them down on the Soyuz, but it still had to be a terrifying moment, wondering how long he'd be stuck, if it was safe to fly back now. And getting those astronauts home was only the immediate problem. Everything about the station—assembling it, supplying it, transporting the crew—was predicated on the shuttle being operational. Now the fate of the whole project had been thrown into question.

The next morning I was on a plane to Phoenix to pick up Rick Husband's mother, Jane, and his younger brother, Keith, to escort them back to Houston for the memorial. I spent a night out there with them, tried to comfort them as best I could, and told them what I knew about the accident, which wasn't much at the time. I flew back to Houston with Rick's family and made sure they were settled in to their hotel. On Tuesday we held the memorial at the Johnson Space Center. The turnout was so big we had to hold it outside on the lawn. So many flowers and wreaths were piled up at the entrance gate that cars could barely drive through it. President Bush came down and spoke, and we had a missing-man formation, where airplanes fly over in formation and one of the jets veers off straight up to the heavens to indicate that you've lost someone. There was a missing-man formation in *The Right Stuff*, too, but that was one part of the movie I would have been happy not to have lived through. After the public memorial came the individual funerals. I went to Mike's and to Rick's. There were so many funerals coming so close together I actually couldn't make it to all of them.

Ilan's funeral was going to be in Israel. A handful of us were asked to go and represent NASA. Steve MacLean was Ilan's CACO. He and his wife went. Carola and I went as well. On the day we arrived in Tel Aviv, a public memorial was held at the Israeli Air Force Base adjacent to Ben Gurion Airport. All the big Israeli politicians were there: Shimon Peres, Ariel Sharon, Benjamin Netanyahu. The next day Ilan was buried in Moshav Nahalal, a small village near Nazareth. The cemetery there is up on a hill, and Ilan's grave was at the top looking out over the valley, a place of honor. The service was conducted by the chief rabbi of Israel, and at the end we walked up and placed stones on Ilan's grave.

Up until that day, Ilan was a great friend and colleague and a guy I played ultimate Frisbee with. It was only when I saw the scale of his memorial and the entire nation in mourning that I fully realized what he meant to the people of Israel. I saw why he'd been chosen. He had all the qualifications as a pilot and an engineer, but he was also someone who represented the best of his countrymen. He had the personality and the temperament to use his position for good.

Ilan's dream was to have peace in his homeland. He always made it clear that he was flying for the state of Israel, not only Jews but Muslims, Christians, everyone. Going to space is one of the few things that unites us as human beings. The Americans, the Russians, the Japanese—once you sign up for this mission, it doesn't matter what flag is on your shoulder. We work together because the goal we're striving for is more important than whatever the politicians are fighting about that week. Before *Columbia* broke apart over East Texas, Ilan was headed home to a hero's welcome. He was going to use that fame to tell the story of space to his own people, to give them a common purpose, something to unite them. When we lost Ilan, we lost more than a great pilot, family member, and friend: We lost someone who could have made a difference in a very difficult part of the world.

After a couple of months, the memorials were over and the public mourning ended. People moved on. But inside the astronaut office nobody moved on. In the weeks, months, and years after, we stayed in touch with the families and loved ones of the crew. Kent Rominger and I would go for T-38 runs up to Amarillo to have a piece of pie with Rick Husband's mom. I would call her every few weeks to check in. Other astronauts did the same with the other families. That sense of brotherhood and togetherness I felt when my father was sick, that really kicked in, especially for the kids. The Girl Scouts' daddy-daughter dance was scheduled a few weeks after the accident. I went with Gabby, but Mike Anderson's daughter suddenly found herself without a dad. Steve Robinson, one of the astronauts in Mike's class, was a forty-something bachelor with no kids. He dressed up in a suit and tie and took her to the dance. I remember telling him, "Steve, it's great of you to do this."

He said, "Are you kidding me? This is the best date I've ever had. I'm honored to be here."

That was the way we all felt. Nothing we did could make up for losing a parent, but we were going the make sure those kids had people there for the school plays and baseball games and birthday parties their mothers and fathers had to miss. Carola and the kids and I stayed especially close to Rona Ramon and her children. Rona stayed in Houston for a couple of years, not wanting to disrupt the kids' lives any more than they already had been, but that meant she was now raising four kids alone in a foreign country. The second-oldest boy, Tal, had his bar mitzvah a couple of months after the crash, and we all went to help out and support him. When Ilan's oldest son, Asaf, won an academic award for being one of the top students in the ninth grade, Rona couldn't make the ceremony, so Carola and I went and watched him get his medal and took him out for sushi afterward. In moments like that we tried to be there and pitch in as much as we could.

After a few years Rona and the kids moved back to Israel. After high school Asaf enrolled in the Israeli Air Force Flight School to become a pilot just like his father. In 2009 he graduated first in his class, just like his father, the first time a father and son had done so in the history of Israel. Flying was in their blood. Asaf said he hoped to become an astronaut one day, too. Then, only a few months later, while flying a training mission in the Hebron Mountains on a Sunday morning in September, Asaf suffered a g-force–induced loss of consciousness in an F-16 at a low altitude. He crashed and was killed instantly. He was twenty-one years old.

I couldn't make it to Asaf's funeral because of my second space flight, but I called Rona. She wasn't doing well. Losing a son was harder than losing a husband. I finally made it over to visit in 2010, and Rona took me to visit Ilan and Asaf in their graves. They're buried side by side in the two plots Ilan and Rona had intended for themselves. Standing over Asaf's headstone, Rona said, "This was supposed to be me."

Asaf's death put an end to flying in the Ramon family. The younger children, Rona grounded them. She went straight to the prime minister and said, "No more." All three of them served, but they never flew. That family had sacrificed enough. David, the youngest boy, still wanted to fly. When she told him he couldn't, he said, "Don't you think I would make a good pilot?"

"Yes," she told him. "You would be a great pilot. Better than your father. Better than your brother. In our family we are very good in the sky, but the sky has not been good to us."

20

WHY WE GO

After burying our friends, one Friday afternoon that June, Digger and I flew down to the Kennedy Space Center to say one last good-bye—to our spaceship. All through February and into the spring, crews were walking the debris field that *Columbia* had left strewn across Texas and Louisiana. Anytime a piece of the shuttle was found, it was collected and shipped to the Kennedy Space Center. There, in one of the hangars, they had an outline of the shuttle on the floor. As the pieces came in they were being cataloged and arranged where they belonged. If they found a piece of the fuselage, it would go here. If they found a piece of the landing gear, it would go there. Like putting together a puzzle. Some pieces were charred and twisted. Others were remarkably intact. I could stand in the middle of it and see: This was the ship that took me to space. My locker was here. The galley was there. That's the window where I listened to Radiohead and looked out at the wonders of the universe. By looking at what survived and what didn't, I could tell where the explosion happened and how the shuttle came apart.

There was a separate room, a private area the public wasn't allowed to see. That was where the crew's personal effects were being
collected. It was surprising some of the things that made it down. A
few photographs survived. Ilan's diary survived. Several of its pages
were readable, too. We recovered parts of their helmets, pieces of
their flight suits. As with the shuttle itself, by looking at what was
left behind I could see how they died.

Digger and I mostly walked around the hangar, not saying
much. That was more or less the mood back in Houston as well.
It was somber. It was like a collective post-traumatic reaction, like
everyone had been punched in the gut. But life doesn't stop. While
the accident investigation board conducted its inquiry, we still had
a lot of work to do. Some of the astronauts were tasked with aid-
ing the investigation. Some of us were tasked with reworking the
shuttle systems to prevent future accidents. Most of us were tasked
with keeping business as usual moving forward, which is what I
was doing as best I could. We still had people on the space station.
Don Pettit and Ken Bowersox ended up staying there an extra three
months while they waited for a ride home with the Russians on the
Soyuz, and I was CAPCOMing for them regularly. Future flight
crews were already assigned, and even with those missions being
delayed and shuffled around, we had to work on the assumption
that the shuttle was coming back online eventually. So we kept
training, kept doing runs in the pool. But there was no joy in any of
it. It felt like everyone was going through the motions. One morning a few months after the accident, I was supporting Scott Para-
zynski on a training run in the pool. We were getting ready to
lower him into the water and he stopped and looked at me. He said,
"Mike, I don't want to do this."

I said, "I don't wanna be here, either, but we have to."

The truth is, for me, things hadn't been going well even before the accident. Sean O'Keefe was sending me out for different media and PR appearances and I was enjoying that part of the job, but my main goal was to get assigned to another flight and get back to space. That had turned out to be a challenge. As soon as STS-109 landed safely, the Hubble team turned its efforts to the final servicing mission, Servicing Mission 4. That last Hubble flight was the flight that every spacewalker wanted. The culture at NASA is about serving the greater good and there is no "I" in team and all that, but people still have egos. People want the chance to tackle the high-profile assignments. People want the chance to do interesting, challenging work. So as the last Hubble flight got under way, there was some political jockeying going on. People were trying to position themselves to get assigned.

On 109, John Grunsfeld had been my mentor and protector. Shortly after we got back, he was called up to DC for a new job, serving as chief science officer. That left me without the advocate I'd come to rely on. When the first set of development runs for the final servicing mission started up I was included in them. Then the second set of development runs started, and I wasn't invited to the planning meetings. Suddenly I was out of the loop.

I went and talked to the head of the EVA branch and asked him why I wasn't being included. He said it was because of my postflight evaluations on 109. They weren't bad, but they weren't great. It was a matter of experience. As far as doing the work and executing my tasks, I'd done fine. But Newman and some of the other people rating my performance said that I hadn't shown the leadership skills to be an EV1. They said I needed more seasoning before I could be the lead spacewalker on a team. Basically, the EVA branch was saying I needed at least one more mission under my belt before taking a lead role. But the EVA plan for the final Hubble flight was the same as it had been for ours, to go out with two leaders and two first-time spacewalkers. I couldn't go back as a rookie, and I hadn't established myself as a leader. I was somewhere in the middle, which meant I wasn't going back to Hubble.

I was disappointed. I'd set my sights on going back. I wanted to be a Hubble guy. Now the discussion around the office was that I'd be better suited to go to station and get more experience there. But there was a long line of people who'd been training on station while I was working on Hubble. Before the accident a whole bunch of station assembly flights were announced, and I wasn't on them. I was in limbo. Then *Columbia* happened and I wasn't sure if I'd ever fly again.

On August 23, 2003, the *Columbia* Accident Investigation Board issued its report. At 81.7 seconds after launch, a briefcase-size piece of insulating foam from the external tank broke off and struck the leading edge of *Columbia*'s left wing and punctured a hole in the reinforced carbon-carbon panels of the wing's thermal protection system. The purpose of the thermal protection system was to protect the shuttle from the 3,000-degree temperatures that are generated by the friction of reentering the Earth's atmosphere, but after the debris strike, the compromised panels allowed superheated air to penetrate the wing, melting its internal structure. The wing sheared off and the shuttle broke apart in the sky.

At the beginning of the investigation, when they first started looking at this foam impact as the cause, I said, "No way. That can't be it." This foam was lighter than air. I could hit you over the head with a piece of it and you wouldn't feel a thing. The truth is that bits of foam had been flying off that tank since the first shuttle flights. The external tank is like a gigantic thermos bottle filled with liquid oxygen and liquid hydrogen, which have to be kept at –297 degrees Fahrenheit and –423 degrees Fahrenheit, respectively. That's what the insulating foam does. For the most part, given the extreme stress of launch, it adhered very well. But around the tank's joints and valves, the foam had a tendency to flake off. The damage to the orbiter was never critical, so it came to be viewed as a maintenance issue, something to be patched up in turnaround for the next launch, and not a safety issue. We looked at the shuttle hitting these bits of foam like an eighteen-wheeler hitting a Styrofoam cooler on the highway. One is going to plow right through the other. The more flights that landed safely, the more that assumption became accepted as fact. But any engineer can tell you: Past success is not an indicator of future safety. The fact that something bad hasn't happened yet doesn't mean the possibility of failure isn't present, and relying on a good track record is no substitute for rigorous scientific testing.

This foam was as light as a feather, which leads one to assume that it's harmless, but low-density objects slow down quickly once they lose propulsion. At the moment the foam broke loose from *Columbia*'s tank, the shuttle was moving at 1,568 miles per hour. The foam hit the wing 0.161 seconds after coming off, but in that microsecond it had slowed down to 1,022 miles per hour, which means the shuttle hit the foam with a relative velocity of over 500 miles per hour. If you run into anything at 500 miles per hour, I don't care how light it is, it's going to do some damage. Which is what the postaccident testing showed. An independent team of investigators took chunks of foam and shot them out of a cannon at the leading edge of a wing salvaged from the space shuttle *Enterprise*, causing everything from small cracks to gaping holes.

Perhaps the saddest part of the *Columbia* tragedy is that we might have been able to do something. While the shuttle was still in orbit, people on the ground were aware that there might be a problem. The debris strike was photographed during ascent and was first noticed by a group of engineers going over the launch footage on day two. From the photos, it was possible to tell that a debris strike had happened, but not how bad the damage was or if there was any damage at all. Since 107 was a science and research flight, it didn't have a robot arm on board to inspect the wing. Discussions were had about doing a space walk to inspect the wing or having a defense department satellite do a flyby to take images, but those were very expensive and risky contingencies that would have disrupted the mission. Because of this blind spot we had—this belief that foam strikes weren't dangerous—an official decision was made not to investigate the extent of the damage.

Another official decision was made as well: not to inform the crew. In the moment the feeling was: *Even if we do find that there's extensive damage to the wing, there's nothing we can do to fix it while they're in orbit. Either the crew risks reentry with the wing as it is or they stay stranded in space until they run out of life support.* In hindsight, the investigation board concluded that if the damage had been identified early enough, another shuttle could have been launched in time to attempt a rescue mission. And if we'd known there was a hole in the wing, there's no way we would have left *Columbia*'s crew there without putting another crew in harm's way to try to save them. But we never reached that point because this collective blind spot kept everyone from seeing how bad the situation might be. Since it was believed nothing could be done about the problem—and since no one could be certain that a problem even existed—it was better that the crew not know. When I spoke to Ilan and Rick and the others on the last night of their mission, they were laughing and excited to see their families again. None of us knew that they were already doomed.

The *Columbia* tragedy is one of those situations where no one person is to blame but ultimately everyone is responsible. We had all allowed ourselves to become complacent about reentry. We were all guilty of underestimating the danger. There were also larger institutional problems with NASA itself, failures of communication, failures of oversight. Like most accidents, *Columbia* was 100 percent preventable. If proper safety protocols had been in place and been followed, our friends might still be alive today.

Before the shuttle could fly again, every one of those problems had to be addressed. We redesigned the external tank so that insulation wouldn't keep flying off it. We improved the imagery on the shuttle for launch, putting high-definition cameras everywhere: on the tank, on the solid rockets, on the ground. That gave us eyes on every inch of the shuttle to see if anything went wrong during launch. We developed tools and techniques to inspect the shuttle for damage once in orbit. Now every flight would carry the robot arm and also a new device to extend the reach of the arm, an inspection boom with high-def cameras, lasers, and other sensors that allowed the crew to survey the entire ship. We developed repair techniques for spacewalkers to go out and repair damaged tiles.

The final component in this new recovery plan was the space station itself. It added another layer of inspections: You could flip the shuttle with the underside facing the station and do a close, thorough scan of every square inch. It's also a safe haven. If you're at station and you encounter a problem you can't fix, the worst-case scenario is you stay there and catch a ride home with the Russians on the Soyuz or you wait for the next shuttle flight to come up and get you.

If the *Columbia* accident exposed NASA's greatest weaknesses, the recovery from *Columbia* may have been NASA's finest hour. No attempt was made to cover up the cause of the accident or deflect accountability. No question went unasked. No assumption went unchallenged. Every single aspect of the shuttle's operations was taken apart, looked at, rethought, and rebuilt. We worked around the clock for two years straight in an all-hands-on-deck effort to understand what had happened and to prevent it from ever happening again. When I look back on it now, it was truly amazing what we accomplished. That superhuman effort was enough to put the shuttle back in operation in order to complete the work of assembling the International Space Station. Which was great, but in the end it wasn't enough to save the shuttle program itself—and it wasn't enough to save Hubble.

On January 14, 2004, President Bush announced what he called a "new vision" for America's space program. That new vision was really a return to the old vision: finishing the space station, building a heavy-lift vehicle capable of taking us out of Earth orbit, back to the moon, and eventually to Mars. But this ambitious, long-term goal would require short-term sacrifice. The money to pay for it would come from retiring the shuttle in 2010 once the assembly of the space station was complete.

In the end, the shuttle was a victim of the compromises that gave birth to it. The shuttle was sold as routine, everyday access to space, but in hindsight that was a bold overstatement. It was always a dangerous, expensive vehicle to fly. Before *Columbia*, we calculated the odds of a total loss of shuttle and crew at about 1 in 150. After *Columbia*, that was revised to about 1 in 75. By contrast, the risk of losing a fighter jet in Vietnam was around 1 in 1,500. The orbiters were getting old and the program was expensive to maintain, and if we kept flying them, another accident was seen as inevitable.

At that point the United States had a huge amount of time, money, and national prestige invested in the International Space Station. We had obligations to our partner countries, which is one reason we stayed committed to its completion. But Bush's speech didn't say one word about Hubble. The rumor around the office was that Washington was going to kill the final servicing mission. Being able to fly the shuttle again was dependent on having a safe haven, which you only had if you went to station. But Hubble was 100 miles higher and on a different orbital path. The Hubble has a low orbital inclination. It flies 28.5 degrees from the equator. The ISS has a high orbital inclination. It flies 51.6 degrees from the equator. It's easy to raise or lower the altitude of your orbit, but it takes an enormous amount of fuel and energy to change the inclination of your orbit in space. It's actually easier to land and take off again at a different angle than to change directions in orbit, which means there's no way to get from Hubble to station. And because it takes more fuel and energy to get up there, you have fewer resources to sustain you if something goes wrong.

At the same time, it seemed inconceivable that we wouldn't go back. First, we still had to address the issue of deorbiting the telescope. The Hubble is the size of a school bus, but it lacks the propulsion necessary to perform a guided entry, meaning it could shower a major city with debris instead of being guided to break up over an ocean. When the Hubble was launched, the idea was that once its mission was over the shuttle would go up and bring it back so we could put it in a museum. Without that, we needed to go back to add some kind of steering mechanism to deorbit it safely.

Second, the amount of new science and research we'd lose without the final mission was staggering. We had two new instruments, the Cosmic Origins Spectrograph and the Wide Field Camera 3, which had already been built at a cost of $200 million. Like all of Hubble's instruments, they were keys to unlocking more secrets of the universe, and we had them ready to go. Now we were going to shove them in a closet and say "Oh, well"? Abandoning the Hubble didn't just mean forgoing future upgrades, either. The gyroscopes were starting to fail again and needed to be replaced or the telescope wouldn't be able to point. Worse than that, the batteries were reaching the end of their life span. We estimated they had about three years left in them. Power is essential. None of the telescope's instruments work without it. Power also keeps the telescope warm. Without heat, the instruments would freeze in the vacuum of space, and if they froze even once, that would be it. We'd be left with a useless hunk of metal orbiting 350 miles above Earth. If we canceled the final servicing mission, we were essentially saying, "It's time to let Hubble die."

One week after Bush's speech, NASA made the announcement: Hubble was off the books. It was too dangerous, too risky. It was a unilateral decision. There was no discussion, no review, no panel to study the pros and cons. The backlash was immediate, and it was *big*. Across the board, nearly everyone in the scientific and aerospace community said it was a mistake. Maryland senator Barbara Mikulski, who represented the Goddard Space Flight Center, went to the press, denouncing the decision and saying she'd do everything she could to reverse it. In the House, Representative Mark Udall of Colorado introduced a bill calling for an independent panel of experts to review the cancellation.

Meanwhile, the team at the Space Telescope Science Institute made their own case for Hubble in the best way they could. They rushed out the release of its latest images: the deepest photographs ever taken of the universe. These pictures showed the most distant galaxies ever recorded, nearly ten thousand of them, some nearly as old as the universe itself. If we wanted to continue to study them and learn more about them, there was only one option: save Hubble.

Inside the astronaut office in Houston, watching these decisions being made up in DC, we were devastated. O'Keefe said he was canceling the flight out of consideration for our safety, but nobody asked us. We were still willing to go. As horrible as *Columbia* had been, the bottom line was that the accident didn't add any new information. It's not like anyone was surprised. We knew the risk was there.

For me, with *Challenger*, the danger had been more of an abstraction. Now it was right in front of me and I wondered how I would react. The truth is, it didn't change anything. Carola and I had exactly one conversation about it. It was maybe a week after the accident. I asked her, "What do you think, now that this has happened?"

She said, "Well, we always knew it could. You've only flown once. Don't you want to go again?"

"Yeah."

That was the end of the discussion. Carola is a physical therapist. Every day she deals with people who headed out their front door and got in horrible car accidents and have to learn to walk again. She knows you can't hide from bad things. You just have to keep doing what you're doing. Astronauts go to space. That's what we do. When Ernest Shackleton set off to cross Antarctica, I'm sure the odds were worse than 1 in 75. That didn't stop him. In the shuttle era, NASA got caught up making nuts-and-bolts justifications about why we go to space when the real answer is just because. We go because we go. We do it because we do it. Because human beings have always done it. It's the reason we first left the caves and poked our heads around the next corner to try to see what the world was about.

Exploration is what we do. It's a basic human need, the drive to know more merely for the sake of knowing it. Understanding what's happening at the other end of the galaxy is a path to understanding ourselves—understanding who we are and why we're here. Five thousand years ago the Earth was small and flat and ruled by angry gods who lived on Mount Olympus. Today the Earth is a giant blue spaceship hurtling through an ever-expanding universe that's 13.8 billion years old.

That's why we go.

The beauty of Hubble is that it is maybe the purest expression of that idea that exists today. Not only is it an instrument that can see farther and deeper into the history of the universe than any other machine ever built, the knowledge that it provides belongs to everyone. The U.S. government has spent billions of dollars on this instrument, and then every year we take the knowledge it provides and we give it away. For free. It's all public domain, and not just for Americans but for everyone in the world. It's done solely for the enrichment of our fellow man, and that's an incredible thing. The need to explore, in its purest sense, is always driven by the desire for knowledge itself, and that principle is so important that people are willing to risk their lives for it. Which is why O'Keefe's announcement about Hubble was such a blow to everyone who cares about the future of space exploration, and it's why we weren't going to let our telescope die without a fight.

In April 2004, about four months after the final servicing mission was canceled, Grunsfeld called me up from DC. He said, "Mass, I'm talking to Goddard about a robot mission to save Hubble." It was a bold idea. If it was too risky to send a crew, why not send a machine to perform the upgrades and repairs, one that could be operated from the ground? We still had the problem of deorbiting the telescope safely. We were going to have to send up a robot to resolve that issue anyway, so why not use that as a pretext to investigate and see if a robot could do more, like swap out batteries and instruments?

Someone in the astronaut office would have to lead the effort, and Grunsfeld told me that person had to be me. He said, "I need someone who knows robotics. I need someone who knows Hubble. Most important, I need someone I can trust. You're the only person who fits that bill." Normally, to start any new program in the astronaut office, you would call the head of the office and he or she would assign someone who was available. Grunsfeld wasn't going to take that chance. He was going outside the normal channels and having a specific request come from the NASA administrator that he wanted me on the project. "That's what's going to happen," Grunsfeld said. "So be ready."

Sure enough, a few days later I ran into Kent Rominger in the hallway and he told me Grunsfeld was putting me on this robot mission. At the time, I was already CAPCOMing, working on contingency shuttle repairs, and doing EVA proficiency training. I asked which of my other duties this was going to replace. "None," he said. "Just add this to your plate. Keep doing everything else, but this is a priority."

I had a hunch what was happening: They were propping open a back door for us to get a manned servicing mission back on the books. I didn't say that out loud, but that's what I thought from the start. The plan in the meantime was to send up a robot arm equipped with special manipulators that could dock to the telescope and perform the repairs. I started putting together a team. My old ice cream–eating friend, Claude Nicollier, the Swiss James

Bond, was a veteran Hubble spacewalker and a specialist in robotics. Ten years after he brought my display project into NASA, I called him up and told him I had a project for him. He signed on right away, and we were working together again. We started doing simulations in Houston, running tests up at Goddard. Some of Claude's European sophistication started to rub off on me. I started drinking a lot of *caffè latte*.

The robot mission was an interesting exercise, and we learned a great deal from it, but the expense eventually killed the project. With all the different contingencies you'd have to plan and design for, it would have cost a bazillion dollars and you still wouldn't get the same quality of repair. Replacing the gyros was too intricate a task for a robot to tackle. Even opening and closing the aft shroud doors was tricky. Ultimately, what the robot mission ended up proving was the value of astronauts. Astronauts can think on the spot, improvise solutions, communicate abstract thoughts. Robots can't. If you design a robot to do A, B, and C, and then you get to space and it turns out the robot needs to do X or Y or Z, you're out of luck. If you have a person with a human brain operating hands with opposable thumbs, you can shift gears on the fly, work the problem, devise a solution. As incredible as the shuttle and the station and the Mars rovers are, the most valuable piece of equipment you can have in space is a person.

In the end, the robot mission did one vital and critical thing: It kept the Hubble servicing team together and moving forward. Gene Kranz told me once that after Apollo was over, to keep his people in one place, he put them at Ellington Field until the shuttle program started. When a project gets dismantled, people disappear. They need jobs. They go work for other companies. They go somewhere and teach. If you let them scatter, you're never going to get everybody back. All those skills, all that knowledge and institutional memory, it's gone forever. If you lose the team, you lose everything. With Hubble, the day after the last flight was canceled, people were already sending out résumés, looking for new jobs. We made sure that as many of us as possible stayed put.

The robot servicing mission brought me back to life. After *Columbia*, for a while the job felt like nothing but death and misery. It took me a year to enjoy being an astronaut again, and the robot mission is what made that happen. My mood changed. I had a challenge, a purpose. I wasn't going through the motions anymore. I started going to meetings at Goddard with the Hubble team. I was doing hands-on testing. I was having fun. Once it turned into a robot mission, the astronauts who wanted the last shot at Hubble and had been jockeying for position dropped out. They didn't want to control a robot from the ground. They wanted to walk in space, so they left to go work on station flights. I was left working as a robot guy, and I was thrilled. It was right up my alley. I knew that telescope backward and forward. I liked working with the team at Goddard solving complex engineering problems. The robot mission also gave me the chance to be a leader in the office, to learn how to manage people. I realized my career wouldn't move forward if I didn't.

A few months into leading that team, what I realized was that, over and above my loyalty to NASA and to the astronaut office, my real loyalty was to Hubble. To Grunsfeld and Nicollier and the other Hubble Jedi in Houston. To Frank Cepollina and Ed Rezac at Goddard. To Ron Sheffield at Lockheed Martin. To Barbara Mikulski and Mark Udall in Congress. They were dedicated people. They shared a passion. They were a team. With each servicing mission the astronauts came in and became a part of the family for a year or two and then moved on, but the team would still be there. I wanted to stay with the team. Even when Hubble was canceled and grounded and off the books, I didn't want to leave. Maybe that's something I learned from forty years of being a New York Mets fan: No matter how bad it gets, you stick with your team and you never give up. We had this extraordinary group of people working together, and we weren't going to let Hubble die. Losing seven close friends on *Columbia* had been a brutal reminder to everyone: You only have one life. You have to spend it doing something that matters. Even if I never flew in space again, if my work on that robot mission in any way helped bring back the Hubble, that would be my thing that mattered. That would be my chapter in the story of space.

The whole time I was working, in the back of my mind I never lost hope that we would get a manned mission back on the books. The status reports I filed on the robot mission, I would always end them by saying, "There is an undercurrent of hope that this will eventually be turned back into a space shuttle mission, but there is not a guarantee." Then, on April 13, 2005, propping open the back door with the robot mission finally paid off. On that day Michael Griffin replaced Sean O'Keefe as NASA administrator. I liked O'Keefe, but he was the first to admit he wasn't a space guy. He was a career politician, a presidential appointee. It wasn't his style to take big gambles. Griffin was a space guy to his core. He was a former chief engineer at NASA, had been the head of the space department at Johns Hopkins University's Applied Physics Laboratory, and had done serious work at different aerospace contractors. He and Grunsfeld had been friends for years, and they both shared a love for Hubble.

Griffin was an outspoken, no-nonsense type of leader. From the second he showed up, it was clear there was a new sheriff in town. On the question of whether NASA should be focused on doing science experiments or exploring space, he didn't leave any doubts about where he stood. In one of his earliest speeches to the troops, I remember him saying, "We're not the National Science Foundation. We're NASA. We go places."

Griffin's long-term goal was to get NASA back on the path of exploration, but he also knew that he was likely to have just three years on the job. A new president was going to be elected in 2008, and a new administrator would be appointed. Griffin wanted something important that could be accomplished during his tenure, something he could point to as his legacy. At that point the space station was close to finished, the shuttle program was already winding down, and the new program Bush had announced was still years away from being operational. So Griffin decided there was one thing he could do to leave a lasting mark. A few weeks after Griffin's swearing in, Grunsfeld was down in Houston and he stopped by my office. "I talked to Griffin," he said. "He wants Hubble. He wants us to find a way to go back."

part

6

Worth the Risk

21

FROM THE ASHES

The first time I met Drew Feustel, I thought, *This guy's gonna be a pain in the ass.* Drew was class of 2000, the Bugs. Shortly after *Columbia*, he was scheduled to go through space walk training, and as part of my new role in the EVA branch I was paired up with him for his first runs in the pool, the same way Steve Smith had been paired up with me. I knew how difficult the training was and I wanted to help the new guy as much as I could. A few days before his first run, I called him up and said, "Hey, you want to go out to the pool and go over some things?"

He kind of blew me off. "Nah, I think I've got it."

Then, the night before the run, I bumped into him in the parking lot at the grocery store. He was sitting in his car, this classic BMW roadster, listening to music with his two boys. I said, "You want to hit the pool early?"

He shrugged. "Eh, okay. Whatever you wanna do."

I said, "Look, this is important. This is your run. I'm only trying to help."

"Yeah, yeah. Okay. I'll be there."

He wasn't there. I went in early the next morning and sat around waiting. Drew showed up late. We rushed through the briefing, and the whole time I was thinking, *This guy's gonna screw everything up. He's gonna be terrible.* Then we got in the water. He was unbelievable. My first time in the water I struggled just to get around. Drew did everything perfectly, and the whole time he was loose and confident—a natural.

Drew Feustel wasn't your typical astronaut. He wasn't a military guy. He grew up outside Detroit and didn't have the grades to get into college, so he went to work as an auto mechanic. It turned out that while he wasn't great at sitting in a classroom, he was a genius with anything mechanical. Give him a lamp or an alarm clock or the engine to an F-16, and he could take it apart and fix it and put it back together better than before. He was a real treadhead. His BMW roadster? He'd rebuilt it himself. While working as a mechanic Drew graduated from community college and then went to Purdue, where he met his wife, Indi, and from there went to Queen's University in Canada, where he got a doctorate in geological sciences.

Eventually I got to know Drew better and I saw that he was cool and laid-back and knew what he needed to do and didn't beat himself up about the small stuff. But that took a while. For a long time after that first run I thought he was just cocky. I didn't know that he was one of the greatest guys I was ever going to meet, or that our friendship would be something I'd need to get through the lowest moment of my entire life.

Once Mike Griffin came on board and gave Grunsfeld the nod to start thinking about Hubble, I started feeling hopeful. There were quiet rumblings around the office: "Hubble's coming back." I was getting calls from the team at Goddard, asking me what I knew and if I was going to be involved. For over a year we'd been using the robot mission to hold the team together. Now there was a sense that we were really going back to work. In July 2005, STS-114 flew, the first flight since *Columbia*. Everyone was anxious. It didn't go well. The shuttle made it home safely, but there was still foam coming off the external fuel tank. We were grounded for another year while they continued to work on the problem. Despite the setback, we kept moving forward with Hubble.

After 109, I was disappointed that I hadn't been upgraded to EV1, and it turned out I wasn't the only one who didn't make the cut. There were six other spacewalkers in my class who did well on their first flights but still weren't reassigned. We needed more EVA leaders. Right before *Columbia*, Dave Wolf was named the new EVA branch chief. He noticed the problem and came to me and wanted to talk about how to fix it. I suggested that we have a program to help junior spacewalkers get upgraded. He liked the idea. "Great," he said. "Why don't you head that up." That was something I learned about the astronaut office over the years: If you propose something, you'd better be ready to run it. Dave gave me an office in the EVA branch and put me in charge of putting the program together and managing it.

Then, in the fall of 2005, Dave Wolf took two months of personal leave and tapped me to fill in for him as temporary head of the EVA branch, which put me in a position to attend the astronaut office staff meetings. At one of these meetings it was announced that Chuck Shaw, a former flight director, would be heading up a panel to explore the possibility of going back to Hubble. The rumblings were now official.

Kent Rominger had taken over for Charlie Precourt as head of the astronaut office shortly before the *Columbia* accident. Rommel was a Navy pilot, a great guy, and a strong leader during some difficult years. He and I had been close friends ever since we served as family escorts together for John Glenn's return flight in 1998. Shaw's committee was going to need a crew representative, and it was Rommel's job to name someone for it. There was a lot of jockeying around that appointment. The people who'd disappeared when Hubble was canceled, suddenly they were back, poking around, offering to help. But Grunsfeld and some of the other Hubble guys went to Rommel and told him, "Mass is the guy you should put on this." I had carried the flag for Hubble when it seemed there was no hope it would ever come back. Rommel recognized that. He put me and Grunsfeld on the panel and said, "Go to the meetings and do everything you can to help them bring the mission back on."

The main question Shaw's panel had to answer was: How do we make sure the crew gets home alive? With the shuttle at Hubble's orbit, we'd have the ability to survey the thermal protection system for damage using the robot arm and the special inspection boom. We'd have the ability to repair that damage up to a certain point, using the different techniques we'd been developing since *Columbia*. What Hubble was missing was the space station, where we had a safe haven in the case of a debris strike we couldn't recover from.

The solution we came up with was a rescue mission, a second shuttle on standby on the launchpad with a crew ready to fly. The only time two shuttles had ever been seen on the launchpad at the same time was in the movie *Armageddon*—and that was make-believe, faked with CGI. We were officially in Hollywood action-movie territory now. If necessary, the two shuttles would rendezvous, payload bay to payload bay, link up with the robot arm, and the crew would spacewalk two by two, translating along the robot arm from one shuttle to the other.

The other question was how long we could keep the crew alive at that altitude to wait for a rescue. "Survival" was a word you heard in those meetings. Shackleton mode. How long the crew could survive came down to how long we could stretch our consumables—food, water, fuel, power. You have only a certain number of days' worth of these, and you want to extend those days as much as possible. We looked at how many systems we could shut down to conserve energy. We'd have to turn off the heat, like in *Apollo 13*. It was going to be cold. What were we going to eat? What's the most nutrient-rich food you can eat without using power to cook? We brought in nutritionists to craft a survival diet. Nuts and protein bars and water—that was basically all we'd have. Since the water we drink is a by-product of the liquid oxygen and liquid hydrogen we burn for fuel, those would run out at the same time. We estimated that if catastrophic damage was found on the first day of the mission, we could survive for twenty-one days while waiting to be rescued. But with every passing day operating at full power, that timeline would contract to closer to eleven days. Once we ran out of power and water, that would be it. The carbon dioxide filters would quit working, and we would slowly asphyxiate—we'd go to sleep and never wake up.

Of course, launching two shuttles would expose two crews to potentially fatal conditions, but at a certain point it came down to probability. What are the odds of taking a hit? What are the odds of that hit being catastrophic and not repairable? Now, what are the odds of that happening twice in a row? Once we did the math on that, the probability got pretty low. It wasn't zero, but it was small enough that we could live with it.

Up at Goddard, Frank Cepollina and the rest of the Hubble team were working on the same mission from the telescope's point of view: What needed to be fixed? What was feasible? What tools would be needed? Since I'd worked so closely with Goddard on the previous Hubble mission, I served as the liaison between the two groups. In Houston, Hubble was one of many programs competing for attention. At Goddard, Hubble was their heart and soul, their bread and butter. They'd never lost hope in it, not for a minute, and they were already itching to get started. Cepi wanted astronauts. He wanted spacewalkers assigned to the mission so Goddard could move ahead with testing and planning. Cepi was the kind of guy who had no problem being aggressive and pushing hard for what he wanted. I'll never forget, he came down to Houston for a meeting with Rommel and told him, "I need a crew."

Rommel said, "I can't give you a crew. You don't have a flight."

And that's the way it was: We were in this strange limbo. We had this huge team of people working around the clock on a mission that didn't officially exist at a time when flight operations hadn't officially resumed. But we never let that deter us. Grunsfeld and Cepi and everybody else, we kept pushing. There was a feeling that we were going to make this flight happen through sheer force of will no matter what obstacles were put in our way.

Late November, we received authorization to start doing development runs in the pool. Rommel needed to put someone in charge of them. Again, Grunsfeld and the older Hubble guys recommended me for it. I'd dreamed of being a Hubble guy, one of the Jedi, and I was honored and humbled that they now saw me as one of their own. Between the robot mission, my work on Shaw's panel, and being the liaison with Goddard, they knew that nobody was as close to that final servicing flight as I was. Rommel called me into his office and told me I was in charge of the runs. Based on the work I'd been doing and the leadership I'd shown in the office, he was confident I could be an EV1.

When I walked out of Rommel's office, I was in shock. To be

a lead spacewalker, and possibly going back to Hubble, was at that point beyond my wildest dreams. It felt more incredible than my becoming an astronaut in the first place. Whenever I'd faced obstacles before—when I'd failed my qualifying exam at MIT or been medically disqualified by the astronaut selection board—there had always been at least a crack in the door for me to fight my way back in. But after 109 and *Columbia*, the idea of going back to Hubble . . . I felt like that door had been closed, nailed shut, and painted over. I was told I wasn't good enough. I was told I wasn't going back to Hubble, that I wasn't qualified to be an EV1. Then *Columbia* happened and I hit rock bottom. But I'd worked hard. My friends, Grunsfeld, and the older Hubble guys had helped me out and kept me going. And now my hopes and dreams had been rekindled. I'd risen from the ashes, and so had NASA.

I couldn't say I was definitely going to have a spot on the Hubble flight, but I got the sense that it was mine to lose. When Rommel put me in charge of the development runs he said, "Assume you're on the flight. Who would you want with you?" I got to pick my dream team, more or less. I brought in several veteran Hubble guys, like Joe Tanner and Rick Linnehan. I also picked a bunch of rookies who hadn't flown but who I liked for the job. Drew Feustel was one of them. Michael Good, whom we called "Bueno," was another. Bueno was an Air Force navigator, a real no-nonsense, by-the-book military type. Feustel was loose and laid-back. Bueno was solid as a rock. They were a good balance.

To fly us on the robot arm in these runs, I brought in Megan McArthur, whom I got to know when I trained her to be a CAPCOM. Most astronauts apply to NASA three or four times and don't get in until their mid- to late thirties. Megan got in on her first try. She was twenty-eight and hadn't even completed her PhD yet. NASA wanted her that badly, and when I met her I understood why. She's one of the smartest, most capable people you'll ever come across, and a terrific person to work with. Any minute with Megan is a fun minute. She became the younger sister I never had.

With those selections, we had one slot left to fill. I wanted Grunsfeld, but Rommel was giving me pushback. Grunsfeld had already flown several times, and Rommel was being pressured to spread things around. He had three other candidates in mind. They were all good astronauts, but they didn't know Hubble. Rommel made his case for assigning them and then asked me what I thought. I said, "If you want to go for a beer, pick one of those guys. If you want to fix that telescope, pick John Grunsfeld." We picked Grunsfeld. John had stuck up for me on 109, and now it was my turn to do the same thing for him. It was the right thing to do and, loyalty aside, he was the right guy for the job.

We did one set of development runs in February 2006 and another in April. In July, STS-121 flew, and it was a great success. The insulation problems with the external tank had been fixed, the inspection protocols worked smoothly, and the crew made it home safe. In September, STS-115 flew, and that flight went perfectly, too. Four years after *Columbia*, the shuttle program was back. The station assembly flights were cleared to resume. Momentum was building. Now we were only waiting for word on when we were going back to Hubble.

In September, Steve Lindsey, who served as commander of STS-121, took over for Rommel as head of the astronaut office. He asked for a meeting with me and Grunsfeld to get an update on the servicing mission. We went into the meeting a bit anxious. At that point, as confident as we felt, we still didn't know if the mission would happen. Spaceflight assignments are always tentative. Things can change, especially when a new chief of the Astronaut Office comes in. Maybe the old boss had a plan, but the new boss wants to throw it out and start over. Even after a flight is announced, you can still be taken off it for various reasons. Nothing is for certain until the rockets light up. That's when you know you're going somewhere.

We told Lindsey about what resources we felt we needed and how we were addressing the risks and gave him the whole rundown. He started talking about the publicity that was going to be involved because of the danger. "There's going to be a big announcement when this comes out," he said. "This is going to be the most dangerous mission in the history of the shuttle program, and the press is going to pick up on that and make a big deal out of it." He said he was telling us so we could prepare our families to start hearing that from the media.

Prepare our families? Grunsfeld and I must have looked a bit confused. At a certain point Lindsey stopped and looked at us and said, "You guys do know you're on it, right?"

After the meeting, Grunsfeld came by my office with this big smile on his face and said, "Do you realize what just happened? He just assigned us to Hubble." For over a year we'd been working on a flight that didn't officially exist. Now we'd been assigned to a flight that hadn't been officially announced. For something like that to happen was completely outside the normal channels, but then there was nothing normal about the way this mission was coming together.

Once full flight operations resumed, things moved fast. Typically the way NASA works is that any major decision takes a couple of months. Hubble came together in a matter of days. On October 26, Chuck Shaw went up to Washington to present his panel's findings to Mike Griffin, outlining the rescue plan and the survival strategy. Grunsfeld was in the meeting. He told me Griffin barely asked any questions at all. He'd already made up his mind and was only waiting for Shaw to confirm what he'd already decided. That was on a Thursday. On Friday an internal announcement was made that Hubble was back on the books. Over the weekend, calls went out to the crew.

Grunsfeld and I didn't get calls. We already knew we were assigned. Bueno and Feustel were assigned to fill out the rest of the EVA team. Scooter was named our commander, which made me feel even better about the mission; I knew Megan McArthur was assigned to be our robot-arm operator, and our pilot was Greg Johnson, call sign "Ray J." Megan called him our "Ray J of Sunshine." He was always happy. Ray J is one of those Navy guys who loves to fly. For years he worked as an instructor out at Ellington, flying T-38s, where he got to love working with astronauts so much he decided to become one. It was a fantastic crew: Every single person was a person I would have picked for the job.

On Tuesday morning, October 31, Halloween, Mike Griffin held a press conference at NASA headquarters to announce the flight: STS-125 would be launching in October 2008 aboard the space shuttle *Atlantis*. The seven of us gathered in Scooter's office on the fifth floor to watch the announcement. Senator Mikulski was there with Griffin. It was a big deal. We were told to be ready for a second press conference that afternoon in Houston. We called down to public affairs to try to dig up some matching NASA polo shirts. Then we went out for Chinese food and tried to process what was going on. We were in a bit of shock.

Normally, to announce a flight, NASA would just send out a press release. Then, once the crew was assigned, another press release would go out a few months later. We typically didn't talk to the media until L-30, thirty days out from launch, and even then the coverage was mostly perfunctory. This was unprecedented. I don't recall any other occasion when the flight and the crew were announced simultaneously with so much fanfare or so much attention given to the crew. Shuttle crews are usually nameless: a team, a unit. But at our press conference that afternoon, all seven of us did a Q&A from the front of the room, and then they broke us out for one-on-one interviews. NASA was really putting us out there. We sat with reporters so long we missed trick-or-treating that night.

The way the media was covering the flight, you'd have thought STS-125 was the last flight of the shuttle era, the grand finale. At that point there were still eighteen station assembly and supply flights left, nine of which would fly after we did, but we probably got more attention than all the others combined. After STS-121 flew in July, *Aviation Week* put the shuttle on its cover. But the headline under the photo wasn't RETURN TO SPACEFLIGHT. It was CLEARING THE PATH TO HUBBLE.

Hubble captured people's imaginations. This flight was important and different, and everyone knew it. We knew the shuttle was going away and this was our last shot at saving the telescope. We were putting our lives on the line to unlock the secrets of the universe. If you're an astronaut, a flight like that is the reason you dreamed of doing the job in the first place. I remember being in the gym and running into Brent Jett, who commanded STS-115 that September. We got to talking about the telescope. "The thing about that Hubble flight," he said, "is that you don't even have to ask yourself whether or not it's worth the risk."

Exactly.

22

ONE LAST JOB

Because this was going to be the final trip to Hubble—ever—we had a long checklist of tasks we needed to accomplish. As always, our first job was to repair and refurbish the telescope's existing equipment in order to keep it working: replace the batteries, replace the gyros. These improvements would take the failing, dying telescope and give it anywhere from five to ten years of new life. We also planned to add a fixture to the bottom of the telescope that would allow an unmanned rocket motor to fly up, dock with the telescope, and guide it down to safely burn up in Earth's atmosphere when it was finally time for the Hubble to retire.

We also planned to give the telescope two major upgrades. The first was to remove the Wide Field and Planetary Camera 2 and replace it with the Wide Field Camera 3. The WFC3 was going to be Hubble's first panchromatic camera, able to observe across the ultraviolet, visible, and infrared spectrums. Young stars and galaxies burn bright in the ultraviolet range, while dying stars and older galaxies emit light only on infrared wavelengths. By spanning that range, the new camera would allow us to observe the evolution of galaxies and see further back in time than ever before. The second was installing the Cosmic Origins Spectrograph, or COS, which was going to measure and study ultraviolet light emanating from faint stars and distant celestial objects, allowing us to study the large-scale structure of the universe and the ways in which planets and stars and galaxies are formed.

A spectrograph works like a prism, breaking light down into its component parts, allowing us to collect data about the object being observed, like its temperature and chemical composition. The public loves the incredible images we get from Hubble's cameras, but for scientists, spectography is a vital part of the telescope's utility. Which is why the single biggest, most worrisome task of the mission was repairing Hubble's other spectrograph, the Space Telescope Imaging Spectrograph.

The STIS was installed on Servicing Mission 2 in 1997 and had stopped working in August 2004 due to a failed component in its low-voltage power supply board. It had been resting in "safe mode" ever since. It was a vital piece of equipment; at the time it failed, it accounted for 30 percent of the research being done with the telescope. The STIS allows us to study the relationship between black holes and their host galaxies. When it was working, the STIS allowed us to examine dying stars to understand what happens to them and why. Most important, it enabled us to study the atmospheres of distant planets with the hope of finding other places in the universe that are capable of sustaining life.

We needed to get the STIS back.

Typically, with Hubble's scientific instruments, we never repaired them. We simply swapped the old for the new. That alone was challenging enough. But we had no replacement unit for the STIS and no budget to build one. We did have room in the budget to try to repair it. There was only one problem: The STIS was never meant to be repaired. The Hubble's instruments were designed to survive the violence of a shuttle launch and the brutal conditions of space. They weren't designed to be opened up. By anyone. Ever. They were sealed up as tightly as possible. Imagine all the great heist movies you've seen, like *Ocean's 11* or *The Italian Job*, where some ragtag crew of misfits gets together to break into an impregnable vault or crack the safe that can never be cracked. That's what repairing the STIS was going to be like.

The power supply we needed to replace was housed behind a panel about 14 by 26 inches. Holding that panel down, in the top left corner, was a metal clamp held in place by two torque-set screws. A handrail that was used for installing and removing the instrument was also blocking the panel. It was held in place by four hex-head screws. The clamp and the handrail and the six screws holding them in place all had to come out.

The chief problem with doing intricate repairs inside the telescope is the risk of foreign object debris: my old nemesis, FOD. The same way you don't want to FOD the jet when you fly a T-38, you cannot allow *anything* to get inside the telescope: a loose screw, a speck of dust, particles of gas. If particles of gas get inside and condense on the Hubble's mirror, they might render the telescope useless. When you're inside that machine, you can't even rub your boots together, because it might give off static electricity. This is why astronauts always compare working on Hubble to performing brain surgery. Once you open the patient up, the tiniest mistake can be fatal.

The screws we needed to remove from the clamp and handrail weren't magnetized, so they weren't going to adhere to the drill bit when we pulled them out; there was a good chance they'd go flying off. Also, each of these six screws had a washer behind its head.

When we pulled the screws out, those washers would float off, too. In addition to the washers, to make sure these bolts never, never, never came loose, the threads of each screw were covered with glue. They were literally glued into place. When we pulled these screws out, microscopic bits of dried glue would come flaking off and float away as well. And since astronauts are working with oven mitts in zero gravity and half the time in complete darkness, grappling with microscopic bits of floating debris is close to impossible.

And we haven't gotten to the hard part yet.

Assuming the clamp and the handrail came off clean, the panel housing the power supply was held in place by 111 very tiny screws. The reason it had so many was to keep the STIS from overheating; each screw acted as a mini-radiator, allowing heat to escape and dissipate into space. Which was a rather brilliant engineering solution to the problem of regulating the instrument's temperature, but it was a design you'd use only if you never planned to open up the instrument ever again. Every one of those 111 very tiny screws also had a washer that could go flying loose, and the threads of every one of those 111 very tiny screws were also glued in place. If that weren't bad enough, 2 of those 111 very tiny screws were covered up by a metal plate. When we move things in space, it's helpful to know where the center of gravity of the object is because that's the point we want to rotate the object around. To help us, the engineers who designed the STIS put a label, a metal plate, marking where the center of gravity on the instrument is. Only now that metal plate wasn't helping us. It was in the way and it had to be sheared off, also without creating any debris.

Then, assuming we could remove the clamp, the handrail, the metal plate, and the 111 very tiny screws, the four sides of the panel were sealed shut by a rubber gasket that had to be peeled loose. Then, once we peeled that loose and made sure no bits of dried rubber gasket were floating around, the panel was still connected to the instrument by a grounding wire. The grounding wire had to be cut without causing any electrical problems. Then, once that wire was cut, we would finally have access to the power supply itself. The power supply was a flat card, about 9 by 14 inches. It looked like any motherboard you'd find inside a computer, and it was held in place by channel locks, also known as launch locks, designed to protect the power supply from the violent shaking that goes along with launch. A launch lock consists of a long screw that, as it's driven in, forces the metal plates holding the board to sandwich together and clamp down tight.

Then, assuming we successfully made it through the clamp, the handrail, the metal plate, 111 very tiny screws, the rubber gasket, the grounding wire, and the channel locks, we still had to remove the failed power supply board, which was seated with a 120-pin connector at the back. We had to slide that board out perfectly straight, making sure that none of these tiny metal pins broke off inside the instrument. Then we had to slide the new power supply in—again, perfectly straight—making sure every single one of the 120 tiny metal pins went in flush. If even one of them bent or broke, the whole repair would be blown. Everything up to that point would have been for nothing.

So how do you crack a safe that can't be cracked when you're 350 miles up in space?

The easiest part, hands down, would be removing the handrail. It took up a single, short entry in our checklist. Line 28: "Disengage handrail fasteners (four)." We were nervous about the 111 very tiny screws, which were hard to work with. But the handrail had four big hex screws with large interfaces, easy to engage. Piece of cake. Nothing to worry about. Removing the clamp would be slightly more complicated but also straightforward. We had a clamp removal tool specially designed to help pry it off and capture the screws as they came loose.

To deal with the 111 very tiny screws we designed a fastener capture plate that would fit with an airtight seal on top of this panel. The plate was made of metal, about a quarter inch thick, with clear plastic compartments that had small holes in them that lined up squarely with each of the very tiny screws. Each hole was big enough to allow a small drill bit through to drive the screw out, but small enough that the screw and the washer and the flaking glue debris couldn't escape. The plan was to go methodically, tiny screw by tiny screw. When we were done removing them the debris would be safely contained. For the two screws covered by the metal label, the fastener capture plate had a built-in blade that, with a turn of a knob, would shear the metal off, exposing the two screws underneath.

The trick would be attaching this capture plate to the panel itself. To do that we had to remove 4 of the 111 very tiny screws from the panel to give us holes where we could drive in guide-posts, stanchions, to bolt the capture plate in place. Since those four screws would be coming off before the capture plate was attached, we needed a way to capture the screw and the washer and the dried glue. Each one had to be removed using a capture bit, a special bit on the end of our power tool that grips the head of the screw with teeth and captures it. Once we pulled the trigger to remove the screw, it would remain attached to the capture bit and we would carefully stow it and move on to the next one.

That still left the issue of the washer, which could still go flying. The plan was to insert a washer retainer, a split ring, that could be pushed over the screwhead and snap into place around it. When the screw came out, the washer retainer would hold the washer in place. Then we would use a washer extraction tool to remove the washer and the washer retainer, clearing the way for me to insert the guide studs for the capture plate. Once the capture plate was attached, we'd drive the remaining 107 very tiny screws, remove the panel, peel off the rubber gasket, cut the grounding wire, swap the old power supply for the new, and finally install a new cover panel. The replacement cover was simple. It worked as a radiator without needing 111 very tiny screws. It simply had two levers to be pushed and locked into place. Thank goodness for small favors.

When the flight was assigned, it was assumed that Grunsfeld would be EV1, leading three space walks, and I would be EV3, leading the other two. Drew was partnered with Grunsfeld and Bueno with me. Per Grunsfeld's standing pronouncement about Hubble, the telescope knew we were coming to fix it, so new things started to break. Three months after we were assigned, the Advanced Camera for Surveys suffered an electronics failure that rendered two of its three channels inoperable. Like the STIS, the ACS needed to be opened up and given a new power supply, something that also was never intended to be done. This repair was at least simpler; accessing the inside of the instrument required the removal of 32 screws instead of 117, using a similar capture plate mechanism. Now that repair had to be developed and planned and added to the EVA schedule as well. After everything shook out, Grunsfeld and Drew were assigned the WFC3 installation, the COS installation, and the ACS repair. Bueno and I were assigned the STIS repair and the replacement of the Rate Sensor Units that housed the six gyros. Each team was assigned to handle one of the battery module replacements and a few other miscellaneous repairs.

I was glad that and Bueno and I got the STIS repair job. I wanted it. I'm not supposed to say that. I'm only supposed to talk about the team and how I was happy to play my part and all that, but if I'm being honest I have to say that I wanted that job as much as I've ever wanted anything. My feeling through all of this, since Grunsfeld tasked me with the robot mission, was that this flight was a date with destiny. All the challenges and obstacles I'd faced in life had brought me here, and the STIS repair—if we pulled it off—would be the most intricate, delicate, and complex task ever undertaken by a spacewalker. It was an opportunity to do something that had never been done in space before.

The STIS repair was such a difficult undertaking that the Hubble engineers built us our own model STIS to practice on in Houston, separate from the replica at Goddard and the mock-up in the pool. We had a special room across the hall from our office that we called the Annex, and we kept the model STIS in there along with space suit gloves and a full set of tools. Every moment of free time we had, Bueno, Drew, and I would be in there practicing. It was my job to do the intricate work of performing the repair while Bueno supported me and Drew ran the checklist. I don't know how many run-throughs we did. We must have done it hundreds of times: Remove the clamp. Drive the guide studs. Remove the handrail. Attach the fastener capture plate. Over and over and over again.

Bueno was my EVA partner, but Drew was the guy in my ear. He and I went through something of a mindmeld. He'd tell me what to do, and I'd do it. It was like we had one brain. We developed our own language. This repair was so intricate and so complicated that every item on our checklist had to be broken down into a half dozen micro-steps that all had to go perfectly. There was zero margin for error. Drew would give me verbal cues from our annotated checklist, and I would execute. Then we'd do it again and again and again. A true friendship developed. I came to realize that I'd been completely wrong about him when we met. His breezy attitude came from his confidence and natural ability, but he was dedicated and conscientious and I knew he'd have my back 100 percent if anything went wrong.

We spent a lot of time talking about what might go wrong. That's true of every flight, but it was especially true of this one. We trained and trained on the new techniques for inspecting and repairing the thermal protection system. We spent hours and hours sitting around a conference room table, doing sims for different contingencies, discussing survival scenarios, how we'd live on nuts and protein bars while waiting for rescue. There were certain failures, like some propellant leaks, that we didn't even bother to sim; there was no way to recover from them if they happened. It was morbid and not at all pleasant, but it served to bond us closer as a crew. We knew we were doing something dangerous, but we were in it together and we'd either make it through as a team or we wouldn't make it through at all.

We also had four rookies on the flight. On 109 we had a great crew, but we had five veterans, some of whom had flown three or four times. Flying is always more fun with rookies because everything is exciting and new. Because of the accident, they'd been waiting for years to get a flight. They knew Hubble was the flight that everyone wanted, but they also knew that, with the shuttle winding down, this might be their first and only trip to space. On my rookie flight I'd assumed I'd be back again and again. This time there was a definite feeling of *Let's appreciate this.* STS-109 was a great flight, but it had been business as usual, astronauts doing what we had been hired to do. We didn't know the shuttle program was going away. We didn't know that this chapter was coming to a close and we needed to savor every second of it. This time, nobody took anything for granted.

Everyone at NASA was behind us, too. Mike Griffin told us if we needed anything—*anything*—we were to call him directly. The support we had from the Hubble team was unbelievable. Whatever we needed, we got. We needed helmet cameras that worked in the pool. We got them. They developed them for us. We needed a lighter power tool. They built us a brand-new one just to shave a pound off the weight. We needed a new mini–power tool to handle those tiny screws. They built us that. All told, we had over a hundred new tools designed and built specifically for the STIS repair alone. We were back in the Apollo days: Spare no expense, make it happen.

Every shuttle crew becomes a family, but that was never more true than with the crew from STS-125. Five of us, Megan and the four spacewalkers, had been together since the start of the development runs. Once we added Scooter and Ray J, the team was complete. We were together day in and day out for over two years, sharing an office, training in the pool, training in the shuttle simulator. We'd drive out to Ellington and get suited up and climb into our T-38s and fly in formation down to the Cape, up to Goddard, out to Ames. NASA sent us to Alaska for a kayaking trip, expedition training, bonding time. We came home from that closer than ever before. We ate dinner together every week. We were a tight group of people who clicked really well.

We knew we'd been given a chance to be a part of something special: the last great flight of the shuttle era. It was a risky, life-and-death mission, but that didn't cast a pall over what we were doing. If anything, it kept us determined to have fun and enjoy every second of every day. There was a feeling that we were on this caper, a great adventure. Some days it felt like we were living in a movie. We were Ocean's Eleven, the Magnificent Seven, the Dirty Dozen—a crack team brought together for one last job, one big score.

The media and the public picked up on the excitement surrounding the flight: We were in a race against time to save the most important scientific achievement of the modern space age. ABC News, the Discovery Channel, producers from *Nova* on PBS, they all asked to shadow us during our training to document the mission. NASA had a relationship with the people at IMAX films going back to the early 1980s and had made several films about the shuttle program, including *The Dream Is Alive*, *Blue Planet*, and *Space Station 3D*, all with footage shot by astronauts on board the shuttle. Once STS-125 was on the books, Grunsfeld started talking with IMAX about following us to make a documentary about the final Hubble rescue mission. They loved the idea. Their camera people started filming us during our training and trained us to use their special cameras in space so that we could document the flight itself.

The media request that turned out to be the biggest deal of all was one that I didn't even understand at the time. This thing called Twitter was just taking off. People were sending out 140-character status updates about what they had for dinner and such. The president had tweeted during his inauguration and apparently it was becoming the thing to do. About a month before we launched, NASA's public affairs office approached me about being the first astronaut on Twitter. They wanted me to send out updates about our training and then send the first tweet from space.

I was happy to give it a try, but I had no idea what it would turn into. I also had no idea what one is supposed to tweet about, so I started telling people what I was doing. On April 3 we were at Kennedy for our Terminal Countdown Test, and I sent my first tweet:

> In Florida, checking out our spaceship "Space
> Shuttle Atlantis."

That was it. When we got back to Houston I started sending out updates a couple times a day:

In a simulator practicing for the first spacewalk
on my mission

In a space shuttle simulation with my crew
practicing our rendezvous with the hubble space
telescope.

Practicing closing the big doors on the hubble
space telescope with spacewalk instructors
Tomas and Christy.

I sent out tweets from the NBL, from the shuttle simulator, from Daniel's Little League games. I started following other people, and they started following me. They had tons of questions, and I tried to answer as many of them as I could. People asked about quarantine, about spacewalking, about our training. Mostly they asked how much I'd be tweeting from orbit: They wanted to share that experience and follow along.

For me, social media changed everything. I could share whatever I wanted with whoever was listening. With every retweet and every answered question, the number of people following me grew—ten thousand, then twenty, then fifty, then one hundred thousand, two hundred thousand. And that was in less than a month, all of them everyday people curious about the behind-the-scenes life of an astronaut. When I climbed into my T-38 and flew in formation down to Florida for launch, I was able to share that with them. When I went through the final fit check with my pressure suit and survival gear, they could be right there in the room with me. And at 2:01 p.m. on May 11, 2009, when *Atlantis*'s engines fired and that giant science fiction monster reached down and grabbed me by the chest and hurled me into space, every single one of them got to come along for the ride.

23

LINE 28

From orbit: Launch was awesome!! I am feeling
great, working hard, & enjoying the magnificent
views, the adventure of a lifetime has begun!

That's what I tweeted once *Atlantis* reached orbit—the first tweet
from space. I continued sending tweets when I could, bringing people along for the journey; but I was busy from the jump, busier than
I had been on 109. I was in charge of the post-insertion checklist,
converting the shuttle from a launch vehicle to a spaceship. Fortunately, I didn't get sick this time and was able to get everything
done. We also had to perform the inspections that, post-*Columbia*,
were now a standard part of our postlaunch procedure. On day
three, Megan successfully grappled the telescope and berthed it in
the payload bay while Grunsfeld, Drew, Bueno, and I inspected our
EVA suits, went over our checklists, and prepared to go outside.

For the first space walk, Grunsfeld and Drew removed the old
Wide Field Planetary Camera 2 and replaced it with Wide Field

Camera 3, equipping the Hubble to take large-scale, detailed photos over a wider range of colors than ever before. They replaced the
Science Instrument Command and Data Handling Unit that had
failed the previous September, restoring the telescope's communication capabilities, and finished up by installing the Soft Capture
Mechanism on the bottom of the telescope.

On the second space walk, Bueno and I had to swap out one
of the failing batteries and install the Rate Sensor Units. While
Bueno was working on those, I started on some get-ahead tasks to
help Grunsfeld and Drew repair the ACS the next day. To get the
camera working, we had to reroute power around it using what we
call a PIE harness, a cable about six feet long. From where we were
positioned to work on the RSUs, I was in a good spot to set this
cable up for the next day's work. I went and retrieved it and hooked
it to my mini-workstation so I'd have it for later.

The next thing I knew, out of the corner of my eye I saw the PIE
harness floating away. Somehow the hook that I'd used to secure
it had come undone and it was drifting off into space. The first
thought to flash through my mind was: *That's the only one we have.*
Sometimes we carry spares, like with the RSUs, but there was no
spare for this harness. If it goes, that's it. There's no fixing the ACS
without it, and we're never coming back here again. I wasn't watching a harness float away—I was watching the future of astronomy
float away.

I was inside the telescope, right next to the star trackers and
the super-delicate instruments we're not supposed to ever bang into
or disturb, but I couldn't let this thing get away. It was already
about five feet above me and going fast. I lunged for it. If I hadn't
been tethered to a handrail, I would have been launching myself
into space, too, never to return. But I knew that I was tethered. I
knew it instinctively thanks to my years of training. I didn't even

double-check before jumping. I leapt up, grabbed the harness, then grabbed my tether and pulled myself back down. Grunsfeld was watching me from inside, and it scared the dickens out of him. He yelled over the comm, "Mass! Watch out!" The whole episode was over in seconds, and everyone at Mission Control was so focused on fixing the RSUs that nobody else took note of the fact that I had nearly sabotaged a key part of the mission.

We completed the RSU swap, and the new battery went in with no problems. The next day Grunsfeld and Drew installed the Cosmic Origins Spectrograph and performed the Advanced Camera for Surveys repair. We were watching that repair closely because it was a dress rehearsal for the STIS repair. If Grunsfeld encountered any problems, he might tell me what kinds of challenges I was going to face tomorrow. But there wasn't a single glitch. Both the COS installation and the ACS repair came off flawlessly. It was as close to a perfect day in space as you can get.

I wanted a perfect day. Every pitcher wants to throw a perfect game at least once. As I sat up that night polishing and buffing my helmet, that's what I was thinking. For five years, ever since the day Grunsfeld called me about a possible robot mission, I'd thought about nothing besides that telescope. Every Sunday at church I'd sit in the pew and think about Hubble. I'd take my kids to a roller rink with their friends, watch them go round and round, and think about Hubble. Now I was getting ready to go out there. I knew this would be my last space walk on the telescope. I figured there was a good chance this would be my last space walk ever. And I was about to undertake the most complex and delicate operation attempted on any space walk ever. This wasn't a run in the pool. Everything had to go right.

We woke up that morning to my song: Billy Joel's "New York State of Mind." Bueno and I started our routine, eating breakfast, putting on our garments, going over our checklists. The whole time I was thinking, *This is it. This is the day.* I knew this day would have a story to it, that it would have a beginning and an end. I didn't know how it would end. I just knew it would be significant one way or another, a day to remember. And it was. Since that day, in every speech I've ever given, I speak about that day.

We were attempting something in space that had never been done before. How were we going to do it? *Could* we do it? The reason we run through these tasks so many times on the ground is not simply to learn how to do the job right but to find out everything that might go wrong. Depending on the complexity of the space walk, so many potential problems can occur. The last thing you want is to encounter a problem you didn't think of or hadn't prepared a solution for. But you inevitably do. Drew and I would talk about this in terms of cars. "There's always something wrong with your car," he'd say. "You just don't know what it is yet." Is there a belt that's about to go bad? A weakness in the front left tire that's about to cause a blowout? You don't know, but the demons are out there, waiting for you. Hopefully they only come at you one at a time.

The Hubble was at the back of the payload bay. I wasn't looking forward to getting back there. As the free floater, going back and forth was always a bit hairy, because you're on the sill of the payload bay, looking over the edge into space, and you feel like you could flip and get out of control and go flying off. You're tethered to the ship, but the length of that tether is fifty-five feet, which is a long way to tumble into space before it catches you. The fear is hard to shake.

Since my space walks on STS-109, one thing had changed. The robot arm had always been berthed on the port side of the shuttle, making it difficult to translate along that path. There are some handrails, but it's tough to get a solid grip in a few places. We'd always gone down the starboard side, which had a clear path with handrails all the way down. Now we had this new inspection boom added after the *Columbia* accident. It was stored on the starboard side, and I couldn't go that way anymore. I had to pick my way along this treacherous path around the base of the robot arm, holding on to a hose here, a screw there, always worried I was going to lose my grip and career out of control. Part of the reason I wanted everything to go perfectly was because I wanted to go out to the telescope once at the beginning and come back once at the end. I wanted to spend the day back in the cocoon of the telescope, where I could concentrate on my job.

Once Bueno and I got to the telescope, Drew started walking us through our checklist and we were knocking off the steps. We were even a bit ahead of schedule. I attached the clamp removal tool. That went fine. I attached the handrail removal tools that would capture those bolts and washers as they came off. That went fine, too. Then I had to remove the four screws that would allow me to drive the guide studs that held the capture plate. These were the four screws that stood the greatest chance of throwing off debris that might FOD the telescope. Slowly, very slowly, I took the drill with the washer-retainer bit and removed the first screw. It came out clean. Then the second. Then the third. As I came to the fourth, I looked at it and thought, *One more of you and I can check you off and never deal with you again.* That's how I felt with every little thing ticked off the checklist: I'm done with that. Never again.

The fourth bolt came out clean. Now, before I could drive the guide studs, I had to remove the handrail. I was using the large pistol-grip power tool I'd used plenty of other times—nothing unfamiliar about it. The two top screws came off, no problem. The bottom left came off, no problem. One more and I was done. I engaged the bottom right screwhead with the tool and pulled the trigger like I'd done a thousand times before. It spun and spun. It gave me a red light. I wasn't getting a good green light. The drill bit was going round and round and nothing was happening. Something was wrong. I looked inside the little window of the handrail removal tool and I didn't see a hex anymore. I saw a deformed, rounded thing that I'd created because I'd stuck the tool in there and pulled the trigger and ground down the screwhead.

Stripping a screw on Earth, while annoying, is not a game-over situation. You just pop down to the hardware store and they've got extractor bits and tools designed to deal with the situation. We were prepared for the small screws to get stripped, but nobody had thought it would happen with the big ones. We didn't have any of the right tools, and the closest hardware store was a long way away.

I stared at what I'd done. *That screw is destroyed,* I thought. *It's never coming out, which means there's no way to get the handrail off, which means there's no way to get the guide studs on, which means there's no way to get the capture plate on, which means there's no way to get the 111 screws out, which means there's no way to get the old power supply out, which means there's no way to get the new power supply in, which means the STIS is broken forever, which means there's no way to discover life on other planets.*

And I'm to blame.

All of this went through my head in a matter of seconds. I looked over at Bueno. He was giving me this wide-eyed look like, *What now?* I actually had the thought, *Hey, maybe Bueno can fix it.* But I knew he couldn't take over the repair. He was my partner, but the damage was already done. I looked back at the cabin. My crewmates were in there, but none of them had space suits on and they couldn't come and bail me out, either. Then I looked down at the Earth. *There are seven billion people down there,* I thought, *and not one of them can help me.* No one can help me. I felt this deep sense of loneliness. And it wasn't a Saturday-afternoon-with-a-book type of alone. It was an alone-in-the-universe type of alone. I felt separated from the Earth. I could see what they would be saying in the science books of the future. This would be my legacy. My children and my grandchildren would read in their classrooms: We might have known if there was life on other planets, but Gabby and Daniel's dad broke the Hubble.

I tried the pistol-grip tool again. I was bearing down hard to try to get traction and catch something to spin this screw. I'm a big guy, and by bearing down that hard what I was also doing was pushing with my feet and putting all that force in the opposite direction into my foot restraint, which was attached to the base of the telescope. There are limits to the amount of force that fixture can take. Looking back on it, I'm surprised I didn't rip the thing right out and put a hole in the side of telescope.

No matter how bad things appear, *you* can always make them worse.

I was making it worse. Drew came over the comm and told me to stop. "Mass," he said. "Don't. Pull. The. Trigger."

We all froze for a moment, not knowing what to do. Hundreds of test runs with an identical screw and an identical power tool had never once resulted in a stripped screw. We discovered later, during an investigation, that the problem was the staking, the glue put on the threads to hold the screws in place. This one screw had more glue on it than the other three. So when we calculated how much torque it was going to take to unscrew the screw from the handrail, we were off. If I'd been using a manual ratchet, I would have felt the extra resistance better and adjusted. But I was using this big, bulky tool set at 60 rpm, the highest setting, and it didn't have as much feedback. Which in hindsight was stupid. There was no reason to do it like that. But everyone was so concerned with the 111 very tiny screws on the panel. Those were the ones we were worried about stripping. The big screws with the large interface, they were nothing. They hadn't even been a topic of conversation.

There's always something wrong with your car. You just don't know what it is yet. The staking on that screw was the demon out there, waiting for me. It was lurking, waiting to be found ever since the STIS was manufactured in the mid-1990s and some technician accidentally put a tiny extra dab of glue on the threads of that screw. I didn't know that at the time. I just felt like I'd messed up. And at that point the origin of the problem didn't matter. Even if it wasn't my fault, it was still my responsibility.

We had a backup plan if we couldn't break torque: Go in with a manual wrench and crank it loose. But we had no plan for a stripped screw. The checklist was useless. Tony Ceccacci, our flight director, was at that moment marshaling everyone in Houston and at Goddard to work the problem. Dan Burbank, our CAPCOM, was relaying their ideas to Drew, who'd then pass them on to me.

The only idea coming up from Houston was to keep trying different tools and drill bits to get that screw out, and the tools they wanted me to try were in the toolbox at the front of the payload bay—all the way at the end of the treacherous path on the port side of the shuttle. Bueno couldn't fly up there on the robot arm. The free floater had to do it—I had to do it. I couldn't say I was scared, but I was. I started picking my way along that sill to get to the toolbox, and over the side of the shuttle I could see the beautiful Earth, only it didn't look beautiful to me. It hadn't changed, but my attitude had. As I made my way up to the toolbox, the doubts and fears that had plagued me for years—doubts and fears that I'd thought I'd put to rest—they all came creeping back. Why did I screw that up? Maybe Grunsfeld should have been doing this and

I should have stuck to the basics. Maybe I wasn't good enough to spacewalk on Hubble in the first place. Maybe I was a bad choice, and that's what the postflight investigation would say: "It was Massimino's fault."

As beautiful as that view was, I didn't care. I didn't care about anything at that point but fixing what was going on. Then I realized that, as bad as I was feeling, being scared and full of doubt wasn't going to help. If I didn't fix this, it was never getting fixed. I got to the toolbox, fetched the tool they wanted me to try, and went all the way back. That didn't work. "Try this one." I went all the way back to the toolbox, fetched the next thing, then came all the way back again. That didn't work, either. Then it was "Try that one," and "Try this other one." I must have gone up and down the sill of that payload bay eight or nine times, fetching different tools. With every pass I lost more hope, and I didn't have much at the start. I knew the repair backward and forward, and I knew there was no way to recover from this. We were grasping at straws. We could keep trying different drill bits, but there was nothing wrong with the bit I was using. The problem was the screwhead. We kept trying things and I kept fetching tools and nothing was working.

I felt like I was living a nightmare. I didn't think we stood a chance. My best and only option was to go out gracefully. I just needed to keep it together and do everything that was being asked of me until we ran out of time, and very soon we were going to run out of time. I'd been going back and forth and trying different tools for over an hour. We were in a night pass when I stripped the screw, and the day-night cycles were passing. I knew that time was ticking down and they couldn't keep us out there forever. Ultimately, it was the flight director's call. I knew he was watching the clock, watching our biometrics, and doing the math. Bueno and I could get more oxygen if we needed it, but our CO_2 filters were filling up. Eventually they'd hit their limit. Typically we plan to be out for six and a half hours max. You can stretch it to seven, but you can't go much longer than that. People start to make mistakes from fatigue. Your life support starts to run out.

At a certain point the question for the flight director becomes: Even if we fix this now, do we have time to finish everything else? Because it's not like we can leave the telescope hanging open all night and pick up and finish in the morning. As long as the answer is "Yes, we can finish," we keep going. As soon as the answer is no, even if we have three hours of life support left, the flight director makes the call, because there's no point in continuing. As the clock ticked down, every minute we were getting closer and closer to Bingo time. I thought we had already reached that point. I was trying everything Dan Burbank was telling me, but my expectation was that Ceccacci was going to tell us to throw in the towel at any moment.

Then Burbank came over the comm and said that they were working on something and I needed to go to the toolbox and get vise grips and tape. Tape? Seriously? I didn't even know we had tape on the shuttle. I thought to myself, *Wow. We are really running out of ideas. We're down to the office supplies now? Are we going to try paper clips next? A three-hole punch?* I translated back up to the toolbox at the front of the payload bay and started digging around for tape. It was dark. I was not happy. I was completely demoralized. At that moment I was as low as I had ever been in my life.

Out of the corner of my eye, I could see Drew trying to get my attention from the flight deck window, maybe ten feet away from me. I didn't want to look up. I didn't want to talk to anyone. I didn't want anyone to see how upset and ashamed I was. Finally I looked up, and Drew had this huge smile, almost like he was laughing. I couldn't say anything, because the ground crew would be able to hear it, so we had to communicate with gestures and facial expressions, like a game of charades. I shot him a look. "What's with you?"

"You're doing great," he mouthed back, giving me a big thumbs-up.

I thought, *What is he talking about? Is there some other space walk going on right now that I'm not aware of? Because the one I'm involved with is a total disaster.* But Drew kept smiling. He started rocking his thumb and pinky finger back and forth, pointing between the two of us as if to say, "It's me and you, buddy. We got this. You're gonna be okay."

If there was ever a time when I needed a friend, that was it. And Drew was right there, just like I'd seen in *The Right Stuff*, the camaraderie of those guys sticking together. I'd been feeling stranded and all alone, but I'd forgotten that my team was right there with me—my crewmates and everyone at NASA on the ground. If this thing kept going south, no one was going to point the finger at me and say, "Massimino did it." We would fail or succeed together, and that's the way it should be. Now, I did not believe Drew for a minute that everything was going to be okay. I still thought all was lost. But I did think, *Hey, if I'm going down, at least I'm going down with my best pals.*

It was at that moment that Burbank radioed in to tell me what was going on with the vise grips and the tape: They wanted me to rip the handrail off. The thought of doing something like that hadn't occurred to me; it ran counter to everything I'd ever been taught about the telescope, which was to treat it as gingerly as possible. But while I was running back and forth like a crazy person trying to fix this thing, Jim Corbo, a Goddard systems manager working out of Houston that day, started wondering if it was possible to yank the thing off. He called Goddard and spoke with James Cooper, the mechanical systems manager for the telescope. It was a Sunday. Only a handful of people were working, but Cooper and Jeff Roddin and Bill Mitchell and the Hubble team up there started running around trying to rig up a test to see if this would work. In less than an hour they had the backup handrail from the clean room hooked up to a torque wrench and a digital fish scale to measure how many pounds of force it would take to break the handrail loose with one screw holding it in.

They did it. Successfully. They called Corbo in Houston with the results. Now Ceccacci and his team had to decide whether or not to give this a shot. If we didn't get the handrail off, the worst case was that the STIS stayed broken but everything else worked fine. But if we yanked this handrail off and debris got loose inside it might FOD the telescope, compromise the mirror. Also, in space, flying shrapnel is a bad idea generally. What if I yanked this thing off and it kicked back and punctured my space suit? Then this might become a matter of life and death.

Ceccacci decided to go for it. It was a gutsy call. But like everything else with Hubble, it was worth the risk. Burbank radioed up and explained it to me. "This was just done," he said, "just now, at Goddard on a flight equipment unit, and it took sixty pounds linear at the top of the handhold to fail the single bolt in the lower right position at the bottom."

Drew said, "Okay. Mass, you copy that? Sixty pounds linear at the top of the handrail to pop off that bottom bolt. I think you've got that in you."

I knew I had it in me. I was a big guy in the best shape of my life. I was nervous about damaging the telescope, but for the first time since the whole problem started, I felt this surge of confidence and hope.

The reason for the tape, I now learned, was to tape up the bottom of the handrail to try to contain any debris that might go flying. I made my way back to the telescope, and Bueno and I taped that handrail up. The whole time Drew and Burbank and I were talking this through. We decided I should start by rocking it back and forth, give it a few tugs to yield the metal a bit, and then give it one clean yank once the metal started to give. If I tried to do it in one go, all that power would be in one motion and it would snap and debris might go flying everywhere.

Right as we got the handrail taped up and were ready to go, Mission Control called up to say they'd lost the downlink from my helmet camera and wouldn't have any video for the next three minutes. I didn't want to waste another second. And if they couldn't see what I was doing, even better. Let's have the party now while Mom and Dad aren't home. "Drew," I said, "I think we should do it now."

He said to go for it. "Just real easy, okay?"

I took a breath, braced my left hand and my feet, and looked at this handrail in front of me. When I was growing up back in Franklin Square, there was one day when I was outside throwing my ball against the front steps, and my uncle Frank came over. This was my uncle who lived across the street. He was covered in oil and grease. My dad came out, they disappeared inside the house for a minute, and then they came back out. My dad had this giant three-foot-long screwdriver with him and he said, "Stop throwing that ball. Come across the street and maybe you'll learn somethin'."

I got up and followed them. Uncle Frank had his car, a 1971 Ford Gran Torino, parked in the street out in front of his house with the hood open. Some mechanic had screwed the oil filter in wrong. Uncle Frank had practically destroyed the thing trying to get it out, and now it was stuck. It was a physics problem, the same problem currently staring me in the face 350 miles above Earth. The amount of torque you can generate is related to the amount of force you apply times the length of the lever; applying force at the end of a long lever gives you more torque than applying the same amount of force on a short lever. So my uncle jammed the end of this long screwdriver under the lip of the filter, wrapped a rag around the handle, and started yanking down on it as hard as he could, grunting and cursing under his breath with each tug: "*Ungh! Ungh! Ungh!*" He did that for nearly a minute and finally the filter broke torque and popped out.

As I looked at that handrail attached to this $100 million instrument inside this $1 billion telescope, after fourteen years of highly specialized training from the most advanced minds in the history of space exploration, all I could picture was my uncle Frank, under the hood of his car, covered in grease, cursing and grunting and yanking on the end of that giant screwdriver. I grabbed the top of the handrail and I rocked it back a couple times and I said to myself, "This one's for you, Uncle Frank." I yanked it hard and *bam!* It came off. Clean. No debris. No punctured space suit.

"Awesome job," Burbank said. "We're back in with the regularly scheduled programming."

Bueno took the handrail, put it in a disposal bag, and we were back in business. I felt like I'd been given a reprieve by God, like I was resurrected from the dead. I felt like this whole episode had been Him giving me a warning to be careful with the rest of the repair. I didn't care what else happened, I was fixing that thing. Nothing in the world—nothing in the universe—could stop me.

I dove straight in and went to work. There was no stopping to celebrate; we weren't even close to being finished. I drove the guide stud anchors. One, two, three, four, they all went in flush. Perfect. I put the capture plate on. It fit, and I cinched it down. Perfect. I took the foil cutter and I sheared off the label exposing the screws underneath. Perfect.

Now I'd reached the big moment: 111 tiny screws and washers to remove without making a single mistake. I grabbed my mini–power tool, I pulled the trigger, and . . . nothing happened. I pulled it again. Still nothing. It was dead. I said, "Aw, for Pete's sake." Bueno and I looked at each other. *What else could go wrong?* Fortunately this wasn't a big deal. Either the battery had died or we'd charged the wrong one the night before, but the spare was in the airlock and I needed to get more oxygen anyway, so it was one more trip across the payload bay for me.

As I was making my way back, two things happened. First, the sun came out. The cold and the darkness had passed and everything was warm and bright and clear again. Second, as I was moving along that edge and looking over the side of the shuttle, I realized that I wasn't scared. I'd been back and forth so many times that this treacherous path wasn't so treacherous anymore. I realized that my doubts and fears had been totally wrong. I was a spacewalker. I *was* the right guy for the job. They *had* picked the right person for this. Because being the right person isn't about being perfect; it's about being able to handle whatever life throws at you. I'd faced every astronaut's worst nightmare, and with the help of my team I pulled myself out of it. And if that problem with the handrail had never happened, I never would have known I had that in me.

I zipped down the side of the shuttle, put a new battery in the mini–power tool, pumped up my oxygen tank, and went back out like a superhero to fix that telescope. And we did it. We hit a couple of bumps, but the rest of the day just went. The screws came out, the panel came off, the old power supply came out, the new power supply went in, and we closed it up.

Once we were done, the team at Goddard performed an aliveness test to see if the STIS was operating again. It was. Everybody started cheering and high-fiving each other, saying "Great job" and "Way to go." I felt a huge weight lifted off my shoulders. Then, while the big celebration was going on, I glanced down at my glove and noticed something: There was a tiny rip in my space suit glove. It was only in the outer fabric. It hadn't gone through the other layers yet, but if we'd seen that rip earlier, that would have been it. Ceccacci would have aborted the EVA and brought us in immediately. The whole space walk would have been over before it started. I suppose that rip was the other demon out there waiting for me. But it didn't get me that day.

After we closed the telescope doors, Bueno was at the back of the payload bay finishing up, and I went back into the airlock to do an inventory and stow things away. Scooter came over the comm.

"Mass, what are you doing?"

"I'm getting the airlock ready."

"Is there anything you're doing now that can't wait?"

"No."

"Well, why don't you go outside and enjoy the view?"

This was the commander ordering me, so I figured I'd better do as I was told.

"Okay."

I went back out, up to the top of the payload bay, clipped my safety tether to a handrail, and I just . . . let go. I stretched out and relaxed, the same way you'd float on your back in the ocean on a warm summer day, and looked at the Earth below. We were coming over Hawaii, a few tiny islands alone in this brilliant expanse of blue. It was beautiful again. Magnificent. I wasn't stealing a glance at the planet while I was supposed to be working, and I wasn't inside, craning my neck to look through a window. I could turn my head in every direction and drink it all in.

We hit Southern California and San Diego, then Las Vegas and Phoenix started whipping by. When I was twenty-one years old, I watched *The Right Stuff* from the balcony of the Floral Park theater on Long Island and saw a sliver of the Earth through the tiny window in John Glenn's capsule. I decided I wouldn't be happy until I saw it for myself, and here I was, except that the view was a thousand times more spectacular than anything he witnessed on that flight. Life doesn't give you many perfect moments. This was one of them. This was my reward, my gift, a few precious minutes to lie back and look at the most perfect, most beautiful thing in the universe. Then, as we came up on the East Coast, I felt that chill in my bones telling me that night was on its way. Out of the corner of my eye I saw a dark line creeping toward me westward across the Atlantic, and I knew it was time to go back in.

It was time to come home.

24

GROUNDED

When you're in space and you want to set something down, like a spoon or a Sharpie, you don't actually put it down. There is no down. You set it out at arm's length and let go and float it there where it'll be handy if you need it again. On my second day home I was unloading the groceries from the store. I grabbed a bag from the back of the car, took it out, stood up, set it out about shoulder high, and let go. It didn't float.

Coming back to Earth is hard. It's an adjustment. After Bueno and I finished the STIS repair, Grunsfeld and Drew had a successful final space walk. The next day we said good-bye to the Hubble, sending it off on its way to unlock the secrets of the universe. We had our normal day off and went through our final inspections. Whenever I could I'd steal a few minutes to go up to the flight deck and look out the window. Outside the window I could see the Ku-band antenna, covered in gold foil, moving and reconfiguring itself. That's how it works: It locks onto the signal from a communications satellite and tracks it to keep us connected to the ground.

Anytime it loses the signal, it pivots and swivels around until it finds the signal again.

Watching the antenna swivel around, with the Earth passing below, I had a feeling I don't think I'd ever had before: satisfaction. I could relax. I was finished. For five years, the Hubble had consumed my every waking moment, and now all that stress and responsibility had floated away. It was a huge relief not to have to think about it anymore. It was done, and I could feel good about it. And I wasn't only satisfied with the mission. My whole life I'd been restless. I always had to do more, reach for the next challenge, the next opportunity. Now I could stop and take a breath. I'd done everything I'd set out to do. Which is a wonderful feeling but also a terrifying one. The signal that I'd been locked onto, the thing that had been guiding me all these years, I was about to lose it, and soon I'd be the one spinning around, searching for the next thing.

The morning we were supposed to fly home, there was bad weather over Florida and we got waved off a day. At that point everything was put away. There was no e-mail. There was literally nothing to do. I grabbed some snacks and my iPod and went right back to the window and I stayed there pretty much all day, listening to Sting and U2 and Radiohead and Coldplay and my Thomas Newman movie soundtracks. The next day we woke up and when we passed over the southeastern United States, the storm clouds were so bad we couldn't see Florida. Sure enough, we were waved off again, and I was back at that window, drinking in the view and savoring every second of it.

Some of the other crew, they were getting bored and started watching movies, ones we had on DVD or that we'd downloaded to our laptops. I skipped the movies. I couldn't imagine tearing myself away from that window for one second to watch something

I could see on the ground. At one point Drew called up from the mid-deck, "Hey! Come down!"

"Why?"

"We're going to watch *Nacho Libre*!"

"I'll see it when we land!"

The next morning Florida was still clouded over, but we couldn't stay any longer. We were diverted to land at Edwards Air Force Base in California. Our families would meet us in Houston. This time I was on the flight deck for the trip home, so I got to see everything: the Earth getting bigger and bigger as we flew lower and lower, the shuttle's nose and tail glowing red-orange hot. There had been an ever-present worry about entry ever since *Columbia*, but as soon as we came out of the darkness over the Pacific and I could see the California coast lit up in the daylight, I knew Scooter was going to get us home safe.

Two days later I was in my driveway, dropping grocery bags and feeling out of sorts. During your first week back, your hand-eye coordination is completely messed up. Your sense of balance is thrown. Your spine is still settling back together, and that can be uncomfortable. You're not supposed to drive or work heavy equipment for three days. Part of it is great, of course: seeing your family, having this wonderful feeling of accomplishment. But then you drive up to the house and real life is there waiting for you: Some shingles over the garage need to be fixed, the pool needs to be cleaned. People always ask me if I miss being in space. "Only when I'm mowing the lawn," I say.

Fortunately for me, even though the flight had ended, the mission was far from over. When we landed at Edwards, one of the administrator's assistants was waiting for us with a copy of the *Washington Post*. There was a big photo, above the fold, of me in my space suit with a big smile in front of the telescope in the payload bay; Megan or John or Scooter had taken it during my last space walk. Hubble was a big story. People wanted to hear about it, learn about it. Because of my experiences and thanks to Twitter, I wound up handling many of the media appearances.

Back in Houston, once my postflight duties were wrapped up, I was offered two different positions. I could be the leader for the incoming class of ASCANs or the astronaut office liaison to the public affairs department. I picked public affairs. I knew my time at NASA would come to a close, and working with media—telling the story of space and documenting the end of the shuttle era—was something I wanted to do. I also knew that, with social media, I could start communicating directly with people all over the world. I checked a camera out from the public affairs office and started taping behind-the-scenes videos with my friends around the office. I knew I could talk to them better than reporters could, get them to relax and have fun. I shadowed them in the NBL, in the shuttle simulator. Together we showed people what the lives of astronauts are really like. Then our public affairs team would edit the interviews and post them to YouTube.

More and more people started following me on Twitter, too. I was lucky to be the first one to use social media, but soon more and more astronauts on the space station were signing on and taking it to new levels. They started making time-lapse videos of the Earth from orbit that, to this day, have been seen millions of times. My buddy Chris Hadfield grabbed a guitar and made a music video, a cover of David Bowie's "Space Oddity" that went viral around the world. Before social media, even when people cared about what astronauts were doing, it was hard to follow along. The Internet's changed everything. We've made the experience of being in space more real for people. They feel connected to what's going on.

Ten months after we flew, IMAX premiered *Hubble 3D*. I was invited on *The Late Show with David Letterman* to promote it. Then the National Geographic Channel saw the videos I'd been posting online and the work I was doing on those other shows and asked me to host *Known Universe*, an eight-part documentary about mankind's quest to understand the cosmos. While I was doing all that, Bert Ulrich, NASA's liaison for film and television collaborations, called me up and said, "Mike, we have an opportunity with this show, *The Big Bang Theory*. You ever hear of it?"

I hadn't. "I knew it was a theory," I said. "I didn't know it was a show."

"It's the number one sitcom in America. It's a science show. They're always referring to NASA and they want to send one of their characters to space and they want to talk to an astronaut. Are you going to be in LA anytime soon?"

It turned out that I was going to be out there for my son's water polo tournament in less than a month. Bert told me all I had to do was drop by the show's production office, hang out in their writer's room, and share some stories. So I did. I met the writers and the producers, Chuck Lorre, Steve Molaro, and Bill Prady. I sat around with them and told them stories for a few hours. This was in the middle of season five. The character they were sending to space was Howard Wolowitz. I flew back to Houston and got a note from Prady a few months later: "Hey, we'd like you to do a cameo." So they wrote me a scene where I give Wolowitz his astronaut nickname, "Froot Loops." I flew out and filmed it. A few months later I was going back to LA to do *The Late Late Show with Craig Ferguson*. Prady called and said, "As long as you're coming out here, we might as well write you into the show again." They wound up writing me into the season five finale and several episodes of season six as Wolowitz's partner on the space station. I've guest-starred six times to date, most recently giving Wolowitz advice on throwing out the first pitch at a baseball game. (Sound familiar?)

The Big Bang Theory was huge. It averaged over 20 million viewers a week, and it will play forever in syndication. It's safe to say more people know me from that than from anything I ever did in orbit. The producers and stars of that show were the nicest, most generous people to work with, and they were so excited to weave their show into the story of space. In the twenty years since the "Deep Space Homer" episode on *The Simpsons*, on the most popular sitcom on TV, NASA went from being the butt of the joke to being the star of the show. Appearing on a sitcom might not seem like a big deal compared to flying 350 miles high in space, but for me it was important. From the beginning, my love of space was shaped by the way astronauts are portrayed in the media and in pop culture. Watching the moon landing with Walter Cronkite, poring through old *Life* magazines, going to the movies to see *The Right Stuff* and *Apollo 13*—those things changed how I felt when I looked up at the sky and dreamed of going there, and I remember that every time I'm given the opportunity to step in front of an audience.

Thanks to *The Big Bang Theory*, I got to go to the space station on TV, but I never managed to make it there in real life. They say you should treat every flight like it's your last. Savor it. At the end of my final space walk on 125, floating above the payload bay, watching the planet pass below, I had a feeling I might never be back. There was a voice in the back of my head saying, *Take a good look, 'cause this is it*. It turned out it was, and that was okay.

For a few months after 125 there was some talk that I might be on one of the final shuttle missions, maybe even the last shuttle mission, but it never materialized. I did have one chance to go back. In April 2010, NASA offered me a spot on a long-duration trip to the station, flying with the Russians on the Soyuz. Being an astronaut is demanding. There's a lot of time away from family, but at least the job itself is in Houston. I was always around for the important things, like birthdays and coaching Little League teams. When I got offered a long-duration flight, I did the math. My daughter was a junior in high school, and my son was a freshman. For the next two years, more than half my time would be spent overseas, mostly in Russia, and then I'd be gone in space for six months after that. If I had never flown in space or if I was less satisfied with what I had gotten to do on my spaceflights, I would have jumped at the opportunity. I calculated the date of my return as the week before Daniel would graduate from high school. I'd miss seeing Gabby off to college. I'd miss everything. It wasn't a hard decision to make.

Passing on a flight is something that's generally not done. At NASA you take what you're assigned. Once you pass on a flight, you're sending a signal to management that your days as an astronaut are numbered. I didn't know it at the time—or maybe I knew it and hadn't accepted it yet—but my time in Houston was winding down. The end was coming quickly now.

One year later, on July 8, 2011, at 11:29 a.m., Gabby and I were standing outside the Saturn V building at Kennedy Space Center in an area called the Banana Creek viewing site for the final flight of the thirty-year shuttle program. The space shuttle *Atlantis* was on launchpad 39A, ready to take the crew of STS-135 up to deliver a year's worth of clothing, food, and equipment to the astronauts on the space station. It was completely surreal that it was happening. When I arrived at NASA in 1996, I figured I'd fly a half dozen times at least. The shuttle program was thriving. The space shuttle *was* space travel. There was no reason to think it would ever come to an end. But it had, and what did that mean? Were we supposed to be celebrating the shuttle's achievements? Sad that it hadn't lived up to its promise? I didn't know what to feel about it. I don't think anyone did.

The launch of the final shuttle mission came with the usual editorials and news segments about why we go to space and whether the expense of the shuttle was worth the return. When *USA Today* published a look back at the shuttle program's triumphs and tragedies, a friend of mine brought me a copy. The tragedies it listed, of course, were the *Challenger* and *Columbia* accidents and the fourteen lives that were lost. As the triumph of the shuttle program, the article cited this:

> On May 17, 2009, floating 353 miles above the surface of the Earth, astronaut Michael Massimino put his gloved hand around a balky handrail obstructing repairs and ripped it off the $1.5 billion Hubble Space Telescope. Only an astronaut could have done this.

Flattered as I was, the point the article was trying to make wasn't really about me. It was about the importance of astronauts. It was the same conclusion we came to after researching the robot mission to Hubble. Unmanned space travel is a great first step; lunar probes and Mars Rovers are excellent tools for scouting a path to explore—but you still need people do to the exploring. What was accomplished on those Hubble servicing missions—upgrading the instruments, repairing the STIS, yanking off that handrail— would have been impossible without astronauts, and we couldn't have done it without the shuttle.

There's an ongoing debate about the most important legacy of the shuttle, whether it's deploying and servicing the Hubble Space Telescope or building and supplying the International Space Station. Whenever I'm asked, I say the greatest thing the shuttle did was that it put a lot of people in space—fifty, sometimes sixty people a year when the program was at its peak. Every person who goes to space, every person who gets to peek around the next corner, is someone with the potential to help change our perspective, change our relationship to the planet, change our understanding of our place in the universe. Which is why we go to space to begin with.

I knew I was never going back, but a few weeks after STS-135 launched, it was made official. I was pulled off active flight status. I didn't want to be taken off. I was sad about it, but the shuttle program was done and I'd made it clear I didn't want to fly on the Soyuz, so a decision was made. I was still an astronaut, but I wasn't going back to space again—and no more hours in the T-38, either. I was grounded. After his one flight on 125, Ray J had left the astronaut office for Ellington, where he became head of flight operations. He called me up one day and told me he'd seen that I was being taken off the flight list at the end of September. "What do you want to do?" he asked.

"I want to fly."

So, for my last couple weeks, Ray J took me flying. We went out and did acrobatics in the practice area over the Gulf of Mexico. We went cloud surfing, did loops and barrel rolls and touch-and-goes on the runway at Ellington. I flew as fast and as high as I'll ever fly again.

My first year as an astronaut, Carola and I both had relatives in for Christmas. I was on the last T-38 flight before the holiday, and I told them, "Hey, I'm flying today. Why don't you come by and I'll show you some airplanes." A bunch of folks came out and I showed them around. Daniel was seventeen months old at the time. I was a brand-new astronaut with this baby boy with curly golden hair. I remember he was wearing this goofy jumpsuit with dinosaurs on it. He was getting the hang of walking and had just started forming real words and he was all over the place, baby talking, "Ba ba ba ba ba," like he was the one giving the tour of the airplanes.

When it was time for everyone to go, I still had to change. Daniel wanted to stay with me, so I said, "I'll take him home." I brought him back to the locker room and he toddled around, getting into everything while I changed. Once I was ready, I called him over and bent down and gave him my little finger. He took it and we walked out together, past all the planes in the hangar, saying "Bye-bye" and "Merry Christmas" to everyone. Then we got in my car and drove home.

For my last flight, Daniel was grown. He was sixteen, almost a man now, his sister away at college. He and Carola and my mom and my sister came out to Ellington for the occasion. It was September 30, 2011, a Friday, the last flight of the day. Ray J took me out for some acrobatics and a quick trip over to Lake Charles. Everyone was going to go out to dinner afterward, but I still had to change and clear out my locker. "Whatever you leave here we're going to throw out," they told me. I said to Daniel, "Why don't you stay and help me clean out my locker? Then I'll drive us to dinner."

Daniel sat with me while I packed up some maps and old boots and a couple of flight suits. There was some chitchat here and there, but it was mostly me and him at the end of it just like we were at the beginning. I closed the locker, spun the combination, and locked it. Right in front of me on the locker was my name tag, MIKE MASSIMINO, JSC, HOUSTON. During my time there'd been so many names on those lockers. John Young. John Glenn. Rick Husband and Ilan Ramon. At one point or another all those names had come off, and now it was time for mine to come off, too. I looked at my name tag and thought, *This is the coolest thing I'll ever do. I got to fly with my heroes, and now it's done.* Then I ripped it off, leaving an empty locker with a strip of Velcro on the front for them to give to the next guy. Then Daniel and I headed out, making the same walk we'd made fifteen years before. Only it wasn't Christmas this time. It was late at night, the sun going down and everybody gone for the day. We walked out through the empty hangar, past the rows of quiet planes, climbed in our car, and went home.

EPILOGUE

Around the Next Corner

When I packed up my locker and walked out of Ellington, I had to face the big question: What do you do when you're an astronaut who can't fly? Having the "astronaut" title on your résumé is a great way to open doors. You can get a job interview just about anywhere, and companies are usually eager to hire you. But it only gets you so far. You've got about two weeks' worth of telling funny space stories before that wears thin and people start to ask, "What else can you do?"

It was a hard transition for me. I'd been dreaming of going to space since I was a kid. I'd never really thought about what I was going to do afterward. Once you leave active flight status, you're supposed to transition to a managerial role or say your good-byes. I didn't want to be management, but I wasn't ready to leave, either. I hung around the office for a while, making my PR videos, doing guest spots on TV, handling whatever I was assigned to handle. I was in denial for a long time. I knew that things were winding down, but I wasn't taking any decisive steps to do something else.

Part of me felt like nothing I did could possibly top what I'd already done. But part of me knew that wasn't true. I did want to do other things, tackle other challenges, but it was hard to admit that to myself.

There came a point, looking around the office, when I knew it was time. Kevin Kregel had retired; he was flying commercial for Southwest Airlines. Scorch had left to fly for FedEx. Digger had been gone since 2004; he moved to Colorado and became a motivational speaker. Scooter had left in 2010 to work for an aerospace technology company. Steve Smith and Rick Linnehan and Nancy Currie were here and there serving in different managerial jobs. John Grunsfeld was leaving for an administrative role at NASA headquarters in DC; I'm pretty sure he'll end up running the whole place someday. Bueno and Drew flew again on station assembly flights and were both still active. Megan was still active, too, but taking time off to have a baby. I was coming up on my fiftieth birthday. Younger people were coming in and stepping up. I needed to make a decision. I thought a lot about the talk Neil Armstrong gave to us my first week at NASA: The important thing in life is having a passion, something you really love doing, and you take joy in the fact that you get to wake up every day and do it.

I began to realize academia made the most sense for me. I'd enjoyed teaching during my time at Rice and at Georgia Tech. In December 2011, Rice University reached out to NASA, looking for someone to be the executive director of the Rice Space Institute. They wanted someone to beef their program up and coordinate its research and activities more closely with the work being done the Johnson Space Center. I applied for the job, got it, and the astronaut office agreed to loan me out; my salary and benefits would still be paid by NASA, but I would work at the university. I led a few seminars, helped to develop the curriculum. It was a great way to ease back into university life. Not long after, my alma mater Columbia started asking if I wanted to come back to the engineering school as a visiting professor, still on loan from NASA. As it happened, Daniel was starting at Columbia as a freshman in the fall of 2013. Gabby was starting her junior year at Sarah Lawrence up in Bronxville. With the kids gone and the shuttle program over, Carola and I didn't have much keeping us in Houston. After fifteen months at Rice, I took the visiting professorship offer and we moved back to New York.

On October 1, 2013, right after we left Houston, the federal government shut down when the House of Representatives tried to use the 2014 appropriations bill as leverage to defund President Obama's Affordable Care Act. NASA was shut down almost entirely; only six hundred of its eighteen thousand employees stayed on to support the astronauts on the space station. Along with most of my colleagues, I was deemed "nonessential" and furloughed without pay for what ended up being two very worrisome weeks. I couldn't believe it was happening. How had we gone from John Glenn being a national hero to a time where astronauts weren't even getting paid?

The government shutdown started, somewhat symbolically, on the fifty-fifth anniversary of NASA's charter. In the Mercury and Apollo years America believed in itself. We pledged our public resources to a lofty common goal and we put a man on the moon as a result—a perfect good. The fact that half a century later our elected representatives were willing to jeopardize that mission over some political squabble says a lot about the faith we put in our public institutions today. We owe it to ourselves to be better.

The year I joined NASA, in the astronaut class of 1996, there were forty-four of us. In the astronaut class of 2013, there were eight. That number should be going in the opposite direction. Our space program is in a period of transition. Some doors are closing and others we're still trying to open. But the difficulty of our present moment should inspire us, not discourage us. There's so much for us to achieve if we decide, as a nation, to commit ourselves to it. The James Webb Space Telescope, Hubble's successor, is set to launch in 2018. Private companies like SpaceX and Virgin Galactic and Blue Origin are putting rockets into space, creating a whole new range of exciting opportunities. The international cooperation behind the space station has put a wealth of resources at our disposal that the Mercury and Apollo teams never had. The Constellation Program announced by President Bush in the wake of *Columbia* was canceled, but it was replaced by the new Space Launch System. Once it's operational it will be the most powerful rocket ever built. It has the potential to give us a permanent presence on the moon, to take us to Mars and back.

Despite the short-term challenges we face, one way or another, I have faith that we'll make it. Humans will never stop going to space. We'll go because we have no choice but to go, because it's what we've always done, since the day we left the caves. I have faith because I've seen the men and women of NASA endure tragedy and adversity and come through it more determined to complete their mission than ever before. I have faith because I see the excitement in the faces of young people every day.

In the fall of 2014, I left NASA to become a full-time professor at Columbia. My main class, and my most popular one, is Introduction to Human Space Flight. The way I see it, I'm training my replacements. My job is to inspire them, to show them what it takes to live and work and accomplish great things under brutally difficult circumstances. I take them from the story of Ernest Shackleton to life on the International Space Station and cover everything in between. Not all of my students will become astronauts—most of them won't—but they may help the environment or cure a disease or create some life-saving technology. The same lessons still apply. I try to teach them to be socially useful, to put their talents in the service of the public good. And I'm not only talking to the students in my class. I travel to high schools across the country, talking to young people by the thousands, encouraging them to go to college, to challenge themselves, to follow their dreams.

When I was living in Atlanta, teaching at Georgia Tech and waiting to see if I was going to be chosen to be an astronaut, the Atlanta Olympics were a couple of months away. They were already running those Olympic commercials on TV nonstop: Coca-Cola and McDonald's and all that. It was a Friday afternoon when I found out that the call from NASA would be coming on Monday. That whole weekend I was up, pacing, anxious. It was everything I could do to get through it in one piece. There was this one commercial I saw that weekend. It had this kid. He starts out a little guy, and he's running around in his yard. Then they cut to the next scene, and he's a bit older, still running. He's on a track now. He's jumping hurdles, and with every hurdle he jumps he's older. He's running for his high school team, his college team. He's training and he's training and then, at the end, he's an Olympian. He's got *U.S.A.* on his chest. He's running hurdles in the Olympics . . . and he wins. He takes the gold. Then, from the finish line, he turns and looks back down the track. He sees the little boy he used to be, standing there looking at him. The whole weekend, that commercial was all I could think about, this kid seeing his dream come true. Was that going to be me?

Still to this day I look back over the obstacles and hurdles I've overcome, and I see that seven-year-old boy standing there with his Astronaut Snoopy in his little spaceman outfit that his mom made for him, and I'm so glad he never gave up. Kids today, they don't have any one big thing like the moon landing to inspire them like I did, but in a way they have something better. They can go on You-Tube and watch Neil Armstrong walk on the moon whenever they want. They can follow the Hubble on Twitter, go online and scroll through mind-blowing images from galaxies millions of light-years away. For years, all I had was a grainy VHS copy of *The Right Stuff.* These kids have the entire universe at their fingertips.

My childhood dream came true, but now I have a new one. I dream that some of these young people, while they're out there clicking around, maybe they'll find out about this book and find a way to get their hands on it—and when they do, they'll know that even if you're a skinny kid from Long Island who's scared of heights, if you dream of walking among the stars you can do it. They'll know that finding a purpose, being dedicated to the service of others and to a calling higher than yourself, that is what's truly important in life. They'll be able to close their eyes and imagine what it's like in space, and when they open them again, they'll look up at the sun and the moon and the Milky Way and see them with the sense of awe and wonder that they deserve.

And those young boys and girls, whatever their space dream is, they'll go for it. Whatever hurdles are in their way, they'll get past them. When they fall down, they'll get back up. They'll keep going and going, working harder and harder and running faster and faster until one day, before they know it, they'll find themselves flying through the air. The hand of a giant science fiction monster will reach down and grab them by the chest and hurl them up and up and up, out to the furthest limits of the human imagination, where they'll take the next giant leap of the greatest adventure mankind has ever known.

ACKNOWLEDGMENTS

Similar to everything else I have done in my life, there is a long list of people who helped me on this journey of writing this book. Special thanks goes to: Tanner Colby, who captured my voice and heart, and helped me to transfer them onto the printed page; Kevin Doughten and the team at Crown Archetype, who saw what this book could be early on and gave me the guidance to make it happen; Peter McGuigan and his team at Foundry, who introduced me to the literary world and led me through the process; and Cathy Frankel and the team at MB Talent Management, for helping me manage all things related to the book. Many thanks also to the people who took time out of their busy schedules to review the manuscript in whole or in part: Megan McArthur Behnken, Fran Massimino, Jessica Marinaccio, Bob Gibson, and Dave Leckrone.

My deepest gratitude also goes to all those people who lived this unlikely journey with me: my friends and family; some mentioned in the book but most not, for lack of pages. You have made my life more than I dreamed of, and I hope this book is at least a small reflection of that.

PHOTOGRAPHY CREDITS

All insert photographs are courtesy of NASA except:

Page 1, top left: Collection of the author

Page 1, top right: Collection of the author

Page 1, bottom left: Collection of the author

Page 2, top left: Carola Pardo

Page 2, top right: Carola Pardo

Page 6, top left: Marc Levine, New York Mets photographer

ABOUT THE AUTHOR

MIKE MASSIMINO served as a NASA astronaut from 1996 to 2014. A veteran of two space flights to the Hubble Space Telescope, Mike and his crews set team records for spacewalking time, and he became the first person to tweet from space. He has played himself on the CBS sitcom *The Big Bang Theory*, was featured in the IMAX film *Hubble 3D*, and has appeared frequently in television documentaries and on late-night talk shows and news programs. A graduate of Columbia University and MIT, Mike currently lives in New York City, where he is a professor at Columbia and an advisor at the Intrepid Sea, Air, and Space Museum.